earthworm

EARTH WORM

THE MOST IMPORTANT ANIMAL IN THE WORLD

GEORGE OLIVER

A DISTANT MIRROR

EARTHWORM
George S. Oliver

Originally published under the title *Friend Earthworm* in 1941.

ISBN 9780648859420

A DISTANT MIRROR
web adistantmirror.com.au
email info@adistantmirror.com.au

Contents

Business is Business

"Business is Business," the Little Man said,
"A battle where everything goes,
Where the only gospel is 'get ahead,'
And never spare friends or foes.
'Slay or be slain,' is the slogan cold.
You must struggle and slash and tear,
For Business is Business, a fight for gold,
Where all that you do is fair!"

"Business is Business," the Big Man said,
"A battle to make of earth
A place to yield us more wine and bread,
More pleasure and joy and mirth.
There are still more bandits and buccaneers
Who are jungle-bred beasts of trade,
But their number dwindles with passing years
And dead is the code they made!"

"Business is Business," the Big Man said,
But it's something that's more, far more;
For it makes sweet gardens of deserts dead,
And cities it built now roar
Where once the deer and the gray wolf ran
From the pioneer's swift advance.
Business is Magic that toils for man.
Business is True Romance.

And those who make it a ruthless fight
Have only themselves to blame
If they feel no whit of the keen delight
In playing the Bigger Game,
The game that calls on the heart and head,
The best of man's strength and nerve.
"Business is Business," the Big Man said,
And that Business is to serve!

—*Berton Braley*

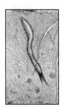

Part I

Introduction

::

There must be something wrong somewhere
— Nature's laws governing these errors — What
is wrong when we have so many human ills?
— When we have want in a land of plenty?
— Why not old age pensions? — Why poultry
dies young — The earthworm as an answer.

There must be something wrong somewhere.

There is want in a land of plenty—nationwide unemployment with factories idle and work to be done. The elderly must depend upon charity or insufficient social security. Fruit trees are chopped into firewood because they die many years before their natural span of life has been completed. Poultrymen and farmers find their livestock dying a premature death by wheelbarrow loads.

There certainly must be something wrong somewhere.

Why should modern man labor under these distressing and unnatural conditions? Are they the fault of a blind and disinterested nature? Or are they of man's own making?

It is the purpose of this book and the lessons it contains—for all books and teachings without purposes are empty things—to encourage the reader to find the something that is wrong and, where possible, alleviate it.

At this point, I wish it definitely understood that I am not coming forward with a new theory, or fad, or panacea for all our individual and collective ills and disorders. What appears in the following chapters was especially prepared for farmers, orchardists, nurserymen, poultrymen, gardeners, trout-farmers and all persons interested in such matters. It is founded on scientifically established data, approved by eminent authorities, and designed by, and functioning through, natural law—the only law from which there is no recourse, no appeal, no hung juries.

Not only are Nature's laws foolproof, but they are as irrefutable and enduring as the laws of mathematics—and two and two will make four as long as numerals are used as media with which to count. And Nature's laws need no policemen, for they are guardians within themselves.

Probably the chief cause of most of humanity's mental and physical disturbances is too much food of the wrong sort, and too little of the beneficial sort. Properly balanced food means a well balanced and healthy body, and such a body contributes to the mental wellness of its owner, or, if you prefer, a spiritual wellness.

When there is want in a land of plenty.

Here indeed is a paradox. If it were humanly possible to keep politicians from signing politically expedient documents and statutes; gag them to silence their similarly expedient utterances and purge their thoughts of forthcoming elections, a fair distribution of the nation's wealth could become a reality. Here, again, we may safely disregard man's laws and turn to natural laws, for an answer.

In Nature there is no waste. Everything—animal and plant, when its life is terminated—returns to its original elements, either in the soil or in the waters. Here, through chemical reactions, it is broken up and again becomes a beneficial part of these elements. Man's laws not only permit waste, but actually seem to thrive upon it.

This is particularly true in political, governmental and social spheres. Modern civilization, of which our American loose-leaf form of Democracy is a major part, seems incapable of producing unselfish statesmen. As examined by the thinkers and philosophers of the world today, general conditions point unmistakably to a decrease of intellectual and moral fiber in those who are elected, or take by force, the responsibility of public affairs. America's financial, industrial and commercial systems need revamping by humanitarians, not politicians. The fact that these systems are all-powerful, as well as gigantic, should not give them a license to act as dictators.

In truth, America has less to fear from a political dictatorship than from a financial, industrial or commercial dictatorship. Certainly no reasonable or reasoning person will deny that our financial system, so cleverly interlocked with the international system, needs a thorough house cleaning. Were this done, and done according to the principles set forth in the Constitution of the United States, industry and commerce would be forced to change their tactics and operate in a less plutocratic, dictatorial and monopolistic manner.

And again we may safely turn to Nature. She does not permit monopoly. Nowhere in either the animal or plant kingdoms will one find monopolistic tendencies. Monopoly, political, industrial or economic, while it is undoubtedly beneficial to a few, is destructive to the masses. In the final analysis, monopoly is self-destructive, and any system that has within it the germ of self-destruction brings widespread disorder to other systems directly or indirectly related to it.

Essentially, monopoly is a form of greed and the similarity is decidedly expressed by calling the reader's attention to a pig pen at feeding time. Invariably, the fattest porker will push and shove and shoulder its way to the feed trough. In its greed, it comes very close to monopolizing all the available food. Thus it grows faster and fatter than the others in the same pen—and reaches the slaughterhouse first! Here we observe how greed and monopoly ultimately leads to destruction.

When old age must depend upon charity or a social security that does not secure.

Do natural laws ordain that the aged, animal or plant should suffer in their senile or infirm years? Definitely, no. Having lived naturally, all animals and plants complete their cycle of life in an even tenor, barring, of course, accidents, which are as much a part of Nature as the ebbing and flowing of the tides. Through progress, enlightenment and education—the latter being far less perfect than our paid scholastics would have us believe—we have graduated from many primitive customs which ordered that the elderly of some races, being senile, and therefore unproductive, should be destroyed.

Yet, the system we boastful Americans employ in caring for our elderly is relatively not far removed from the barbarous system of killing the old. Today we do not kill our elderly outright. We condescend to permit them to slowly starve or freeze to death, making them all the while more susceptible to the ills that come with old age.

Many Americans make the elderly a political football to be booted about the political gridiron. In considering the problem of our elderly, politicians obviously refuse to consider Nature. Probably, few of them realize that mutual aid is a natural law; that very early in the nineteenth century, philosophers dimly developed the theory that in every branch of Nature mutual aid is as permanently fixed as the laws of conception and demise.

It was Professor Kessler who, in 1880, while Dean of the St. Petersburg (now Leningrad) University, declared reciprocity to be a natural law. Prince Peter Alexevitch Kropotkin, Russian geographer, having absorbed Professor Kessler's declaration that mutual aid in Nature was irrefutable, wrote and published (1902) his great work *Mutual Aid as a Law of Nature and a Factor of Evolution.*

What both of these men know, or should know, is that the soil is deficient in one or more vitally necessary ways or elements. The orchardist who begins "guessing" about what is wrong with his soil is playing an ultimately losing game. The law of averages is against him. Such a man, desiring to learn the true nature of his soil, the element or elements it lacks or with which it may be over supplied, should have samples of his soil analyzed by capable chemists. Most soils are deficient in elements necessary for plant life not because the elements are not present, but because they are unavailable to the plant roots.

It is into this picture that the burrowing earthworm makes its appearance. All the elements that are in the soil, but which are hidden and unavailable to the plant roots, are broken down by the earthworm and made available. Man has yet to invent, devise or manufacture any machine, any solid or liquid fertilizer as efficient as the earthworm. In this invertebrate animal, Nature has a perpetual soil builder, a four-in-one creature that acts upon the soil as chemist, triturator, cultivator and distributor—as shall be seen.

When poultrymen find their chickens dying a premature death by wheel barrow loads...

For over a quarter of a century, Southern California has been a mecca for thousands of individuals and families trekking here to enter the poultry business. Scattered, estimated and authentic figures give us the information that, in a period of fifteen years, well over 50,000 such business ventures failed. Among those who have managed to remain in business are many who suffer a poultry mortality that is astounding. One set of figures shows that some sixty percent of pullets die before they reach maturity. Others show that fifty-five percent of laying hens have to be replaced every second year, when they should live and be productive for from four to six years. Still other data point out that it costs many poultrymen twenty cents a dozen to produce eggs, when thinking members of their business are producing the finest eggs for less than ten cents a dozen. Poultry raisers and breeders of fowls for meat seemingly have great difficulty in producing birds with a necessary amount of feathers on them.

Why do such conditions exist in the poultry business? When we come to the portion of this work that deals with poultry, we shall learn that the cause of all these conditions is in the poultryman himself. It is not the nature of domesticated fowls, nor is it Nature's design, to suffer such a high mortality rate, produce sterile eggs, or half bald chickens. For over forty years, this writer has known and demonstrated the fact that, if there is little or no deficiency in the diet poultrymen feed their flocks, there will be a minimum of premature deaths, unfertile eggs, and featherless birds. It is no idle remark, no promise of magical prowess, no guesswork to declare that all of these adverse conditions may easily be remedied by the proper application of the earthworm, as we shall eventually see in a chapter devoted to this subject.

The statement taken for our premise—"there must be something wrong somewhere"—may be accurately changed to "there *is* something wrong somewhere."

It is the aim of this book to point out what and where that wrong is and how it can be overcome. Around and upon one word—earthworm—is built the highway to better and more productive trees, plants, vegetables, poultry, game birds and fish.

Such an animal as the earthworm, whose importance is universally accepted and admitted by scientists, deserves a more pretentious volume than this humble effort of mine. But I find solace and satisfaction in having prepared a book for the lay reader in which I have eliminated, as much as possible, the use of confusing technical and long, jaw-breaking zoological terms.

In preparing this work, I have included data about the earthworm that has long been recognized and admitted by men and women of science. In addition, I have included many facts which I have discovered through nearly half a century of experimentation.

This work was planned to be of especial interest to farmers, orchardists, nurserymen, gardeners, poultrymen and all others interested in agriculture, horticulture and their kindred professions. However, it will, I hope, be welcomed as an instructive review of the life and habits of the annelids discussed herein.

In offering this work to the public, I sincerely hope that it will add, to an already long list, many new and appreciative admirers of our friend, the earthworm.

—*George Sheffield Oliver*

History of the Earthworm

::

The animal kingdom — Earthworm low in animal life — Description of various types of worms — Charles Darwin's opinion — What is "Dry land?" — External description of the earthworm — Internal description — Its sexual life — The eggs of earthworms — Their progeny

The animal kingdom is divided into two subkingdoms, invertebrate and vertebrate animals—animals with backbones and animals without backbones. The invertebrate group is distinguished by nine phyla, or divisions. In this group there are over 500,000 known kinds of animals, ranging from the lowest form of animal life, minute single-celled protozoa, to arthropoda—crabs, insects and spiders. In the vertebrate group there are well over 30,000 known kinds—fishes, amphibians, reptiles, birds and mammals.

When it is stated that in this vast array of creatures the lowly, segmented earthworm is probably the most important to mankind, most may find that illogical and unreasonable. Yet, few creatures equal the burrowing earthworm as being essential to better health and greater growth to plant and vegetable life. Therefore, indirectly, it is of the utmost importance to man.

The burrowing earthworm is Nature's plow, chemist, cultivator, fertilizer, distributor of plant food. In every way, the earthworm surpasses anything man has yet invented to plow, cultivate or fertilize the soil.

While it is unquestionably true that plants and vegetables grow and reproduce their kind without the aid of the earthworm, most naturalists claim that all fertile areas have, at one time or another, passed through the bodies of earthworms.

It is also true that the finest plants and vegetables become healthier and more productive through the activities of this lowly animal, which the ordinary person considers useful only as bird food or fish bait.

The earthworm has been playing a very important role in the drama of plant life from time so distant that scientists can merely guess as to the age of this invertebrate animal. Regardless, scientific men are agreed that mankind may rightly acknowledge the earthworm as one of his best friends.

In this chapter, or lesson, the reader will be presented with a brief genealogical background of the earthworm and the manner in which it has indirectly aided mankind by directly aiding plant life. This background should help the reader to understand facts regarding the earthworm which should be known to anyone interested in gardening, farming, orcharding or poultry raising.

If must first be realized that there are worms and "worms." All are invertebrate animals. This work shall be focused on only the *phylum annelida*.

The division of invertebrate animals, of which the earthworm is a member, is composed of five families or classes. These, in turn, are divided into two orders. The *phylum annelida*, the entire division of earthworms, contains upward of eleven hundred species.

Of this extensive array, we shall concern ourselves only with earthworms, for there are marine worms, swamp worms and beach worms, many of which appear to be "just worms."

While all *annelida* are, more or less, closely related, each specie has distinct features. Some have habits quite foreign to other species. Some prosper only in certain, specific environments and die if transplanted elsewhere. Some have definitely formed heads, with whiskers, teeth and eyes. Others have no heads, are toothless and eyeless. Some worms are hermaphroditical, others bisexual. Some live exclusively in water, others in soggy soil, others in decayed animal matter (manure), others in decayed vegetable matter (humus).

Low as earthworms are in the scale of life, they show unmistakable signs of intelligence. Charles Darwin's experimentations with them conclusively proved that instinct alone could not guide them so consistently. (See Darwin's famous work, *The Formation of Vegetable Mould Through the Actions of Worms, with Observations on Their Habits*.)

Some earthworms come to the surface of the soil and can crawl a great distance, especially in rainy weather, when their burrows or tunnels are flooded. All throw their bodily excrements, technically known as castings, behind them. Some species throw their castings above the surface of the soil, forming small hillocks or mounds.

Countless thousands of years before the rocky surface of the earth disintegrated to form what we call soil, an extensive list of animals and plants lived in the waters. Marine worms were undoubtedly present in those obscure ages.

In time, as the waters receded, various animals and plants evolved certain anatomical organs to meet the new conditions. Some marine worms acquired physical characteristics which permitted them to live, first in very marshy ground, later in "dry land."

The phrase, "dry land," should here be qualified, for, in the strictest sense, there are exceptionally few spots on the face of the earth that are dry. No creature can live on, or in, dry land. It is a common remark, "we breathe air," but what we are actually doing—what all living things are doing—is breathing nitrogen dissolved in water.

We should keep this fact regarding the vital need of water constantly before us as we study the worm and its relation to plant life, for both must have moisture to live.

Now that we have cursorily traced the earthworm from its parent environment to the so-called dry land, we will focus on those known to science as *Oligochaeta*.

This group is composed chiefly of terrestrial worms, and is the subject of this book.

The earthworms, like all other families, is sub-divided into various groups, but for our purpose all we need know are the common names for this class. These are orchard worm, rain worm, angle worm, dew worm, brandling, compost worm, night crawler, fish worm, night lions and similarly descriptive names familiar to certain areas of the United States.

Let us now combine all these common names and visualize the last earthworm we saw.

In size, it may have been from two inches to perhaps a foot in length. Although, twelve inches is long for an earthworm on the North American Continent, except in very damp forest lands.

In considering an external description of the earthworm, we find all species so much alike that few can distinguish one species from another without careful examination.

All are "headless," eyeless and toothless. There are no external antennae or feelers. From tip to tail the body is composed of ringlike segments. A short distance from the "head" is seen a band, which is lighter in color than the rest of the body.

That, briefly, describes how the earthworm appears to the naked eye. The earthworm's internal system is highly complicated. Yet, paradoxically enough, it is magnificently simple. Picture a flexible metal tube the size of a lead pencil, in which is built a plant capable of refining gasoline from crude oil. In a comparative sense, the earthworm's system does to soil what the modern refinery does to crude oil.

The earthworm has a multiple system of hearts, minute tubes circling that part of the alimentary canal between the pharynx and the crop. Through a complicated system, these hearts supply blood to all parts of the body.

Minus lungs, the earthworm "breathes" through its moist epidermis or outer skin. The blood corpuscles are colorless and float to the surface of each segment where they absorb oxygen.

Under an ordinary magnifying glass, the pores of the various segments are visible. If one were to gently squeeze an earthworm, minute drops of yellowish serum would be seen coming out from it.

This serum is composed chiefly of oil of high medicinal value. Experiments for its extraction, discussed in a later chapter, are now in progress. It is hoped that this oil may be extracted in quantities sufficient to encourage production.

Except for a number of hearts, all the vital organs of the earthworm are under the previously mentioned band, which zoologists call the *clitellum*. This band is the chief characteristic of the earthworm, distinguishing it from all other worms except a few leeches and a few other marine worms.

Under this band, in compact uniformity, are seminal vesicles and receptacles, testes, ovaries, oviduct and egg sac. Directly behind these is the crop, where the food is held until the gizzard, just beyond the crop, is ready to accept it. Next follows the intestine, a distinctly oval shaped tube, and then the rest of the alimentary canal to the vent or anus.

Our earthworm is bisexual, containing both male and female organs of procreation, and must perform a reciprocal act of copulation to fertilize and be fertilized.

The sexual act of the earthworm, usually occurring in the cool hours of the early dawn and twilight, makes an interesting and curious study of nature's method for propagating the specie.

Neither animal has external sexual organs, though the pores, through which the seminal fluids appear, are visible under a small magnifying glass. The sexual act is not preceded by any display of amorous cooing or lovemaking. The worms, driven solely by instinct when the procreative glands demand relief, seek a position that brings their bands together and remain thus, quite motionless, for as long as fifteen minutes. If exposed to a bright light during the sexual act, the embrace is broken. Worms, though sightless, are very susceptible to light.

During the act of coitus, each worm exchanges male sperm, impregnating, or, at least, theoretically impregnating, their female ovas. Also during the act, there is an increased flow of the fluid which keeps the entire length of the worm's body moist. This fluid forms the capsule in which the eggs are deposited. It is heavier and thickens rapidly.

When the hymeneal act is completed and the earthworms separate, this fluid forms an outer band. The new band or shield begins to move forward, eventually dropping from the earthworm's "head."

During the forward movement of the gelatine-like band, the impregnated eggs are held firmly within. As it drops off the earthworm, it closes into a yellowish-green pellet or capsule, slightly larger than a grain of rice. This capsule resembles, to a remarkable degree, a very small currant.

Earthworm capsules examined under a powerful microscope show a lack of uniformity in the number of cells. There will be, however, from three to fifteen fertile eggs in a capsule.

Earthworm eggs hatch in about twenty-one days. The newborn appear as short bits of whitish thread about one-quarter of an inch in length. In twelve to forty-eight hours, they become darker but are visible to the untrained eye only after a painstaking search for them.

Once hatched, it is a case of each worm for itself. Close observation seems to lead students of these lowly organized creatures to believe their mortality rate exceptionally low.

Worms begin to mate from sixty to one hundred days after birth, depending upon the richness or poorness of the soil in which they live or in which they are cultured.

Mating follows at periods from six to eight days. If we are to follow the average fertility of each capsule laid, that is, three worms, one mature worm will beget over one hundred and fifty worms each year of its life. Each mating, should produce twice that number, or over three hundred worms a year.

Certain species of earthworms, particularly those that come to the surface and crawl about during wet or rainy weather, seem to be chiefly active during the nocturnal hours. Other species—which we will discuss later—are, apparently, active throughout most of the day and night. This specie seldom, if ever, comes to the surface, depending on the porosity of the soil.

Except in highly porous soils, the earthworm must eat its way through. Having no teeth, everything before it, if not too large to swallow, is sucked into the mouth. It is a ravenous eater.

Every morsel of soil and decayed vegetable and animal matter taken in by the earthworm passes through its digestive system. This is equipped with a gizzard-like organ. Here the food value in the swallowed matter is extracted for use by the worm. The rest is carried by muscular action down through, and out of, the alimentary canal. This waste matter is called castings.

The Habits of
the Earthworm

::

*Terrestrial earthworms differ from other
annelids — Are found in nearly all parts
of the earth — Man helped to scatter them
— How the compost worm lives — How the
orchard worm lives — Habits of both compost
and orchard worms — Diet of earthworms
— How one man made a mistake in feeding
earthworms — Nature's scheme — Man can
improve upon Nature*

As we have already discussed, the burrowing earthworm of our time is an animal that evolved from similar animals which once lived exclusively in the waters of the earth. While the terrestrial earthworms differ greatly from their marine relations of today, there are, many features and characteristics in both that are relatively alike. It is on these likenesses that science bases its contention that the earthworm evolved from its marine prototype.

Earthworms abound in practically every geological section of this planet. The exceptions to this rule are the extreme northern and southern latitudes where extended cold periods do not allow for the existence of this branch of invertebrate animals. But in torrid and temperate zones, some one thousand species of earthworms live, prosper and procreate.

The dense, humid jungles near and around the equator give us the largest specimens of the earthworm. These are undoubtedly the direct antecedents of all terrestrial worms that have spread from one end of the earth to the other.

Assuming that the first progenitors of our present-day earthworm began near the equator, it is puzzling how they have become so widespread between the two frigid zones of the earth's surface.

This is satisfactorily explained by the realization that many species of earthworms peregrinate—they travel and migrate extensively. Some species are known to scale and cross high mountain ranges, though such migration probably required many hundreds of years.

Man, too, has undoubtedly, though unknowingly, aided in transporting earthworms from one hemisphere to another. This has been accomplished through the movement of trees and plants in whose roots worms, or their eggs, have been hidden.

It is quite possible that the early species of terrestrial worms were habitués of soil rich in organic matter or humus. It is also quite possible that these earthworms lived exclusively on the diet supplied them through humus and that the common brandling, or compost worm, is an evolutionary product of them.

In the early works regarding the life and habits of the earthworm, we find no references to the compost worm. All references to earthworms lead us to believe that early students of this branch of zoology, though they examined and studied the digestive organs of their subiects, failed to discover or to realize, or failed to describe, how easily the earthworm becomes a slave to its environment.

The compost worm demonstrates this slaveship admirably. As its name implies, it is found exclusively in manure compost piles, or in soil heavily laden with decayed animal matter. Tests have proved that in such fertilized soil the compost worm will become extinct if fresh manure is not repeatedly added.

To understand the peculiar dietary demand of the earthworm is vitally important for anyone interested in the habits and life of the earthworm. Without this knowledge, any attempt to domesticate the earthworm for fertilizing purposes will be fruitless.

The compost-raised earthworm cannot absorb nourishment from soil which does not contain decayed animal matter. Because of this, any transplantation to ordinary soil would be fatal. Likewise, an attempt to transplant an orchard worm to a compost pile will result in its death.

Strangely enough, if the eggs of the compost worm are gathered and placed in a rich soil minus decayed animal matter, a large percentage will hatch and prosper. While the first offspring may not be high in quality, either in size or health, they will eventually become accustomed to their environment. Each following generation will show a decided improvement. It can be concluded from this premise that the brandling or compost worm is a distinct specie of earthworm.

The brandling, or compost worm, is almost one-third smaller than the common orchard or rainworm. But being of the same family, its head band and rings are identical, except that they are more emphasized.

The compost worm seldom burrows deeply beneath the surface. This is probably due to the fact that soil fertilized with decayed animal matter seldom goes deeper than eight to twelve inches. And, since it is from this organic substance the brandling gets its food supply, it remains within that depth.

Another characteristic of the compost worm is the fact that it does not throw its castings above the surface. Because it is a much fatter worm than the orchard worm, it can release its excretions behind it without fear of packing the tunnel through which it has eaten its way.

These conclusions have been reached after more than a decade of close scrutiny of the habits of the earthworm. The reader may prove these facts to his own satisfaction, and there is considerable evidence available to further substantiate these conclusions.

Like every other animal, the earthworm receives certain dietary necessities from what it eats. If the soil it lives in does not contain these life-giving essentials, the worm suffers. Such deficiency in quality of food will ultimately cause death.

All earthworms eat raw and cooked meat, seldom anything putrid. They like fats, nuts, milk—in short, anything and everything that enriches the soil.

The digestive fluid of the earthworm is of the same chemical nature as the pancreatic secretion in higher animals. This accounts for the worm's ability to digest meats and fats as well as starches and sugars.

To illustrate the manner in which the dietary habits of the earthworm are governed by their environment, the following actual story is worthy of consideration here.

A California orchardist developed a culture bed of 50,000 breeding prolific earthworms. A later chapter is dedicated to their history and development.

He had been instructed how to properly feed his earthworms, but discovered that he could procure beef suet at an invitingly low price.

Believing that he could revolutionize his system of feeding, the orchardist began to place small pieces of suet in his culture trays. Slowly, he increased the amount until eventually suet became the worms' complete menu.

This method of feeding reduced the labor in tending to them, but the day of reckoning was not far distant.

When he was ready to harvest his crop of egg capsules, he discovered that the breeders had failed to breed. What few eggs there were contained no live germs.

He thought to place the obviously sterile worms within some of his trees. The worms' digestive organs had adapted to an all-animal fat diet by slow degrees. They could not adjust to an organic diet in time to prevent starving to death—which they did in a short time.

Similar experiences are on record, and the lessons learned have led us to the logical conclusion that the soil should be impregnated, not with young or mature worms, but with their eggs.

By this method, tests have proved that when the worms are

hatched in the environment in which they are intended to work, they will adapt to its food.

From this brief description of the dietary habits of the earthworm, we come face to face with natural laws reminding us that Nature constantly works blindly. She adheres to only one rule—the continuation of her various species—and is not interested in improving individuals of any of her species.

This fact has long since been accepted by man.

One should not be considered egotistical for claiming that he can improve upon Nature. Every stock farm and nursery disproves the once accepted statement, "You can't improve upon Nature." Our finest horses, cattle, dogs and other domesticated animals and fowls, as well as many trees, plants and vegetables, are the result of man's persistent and intelligent efforts to improve what Nature has given.

If these man-bred animals, trees and plants are not carefully cared for and properly mated by man, they will eventually become atavistic.

They will revert to their ancestral type—a type far inferior to the product evolved by man's intelligence.

Therefore, we should accept the earthworm as an important part of nature—as our friend, a natural friend that should be cultivated, developed and domesticated. By encouraging the earthworm to do its part, we will benefit.

> *"The possibilities of thought training are infinite, its consequence eternal, and yet few take the pains to direct their thinking into channels that will do them good, but instead leave it all to chance."*
> —*Marden.*

Habits of the Newly Developed Earthworm

⠿

Why Nature's earthworm will not function — How the newly developed, prolific hybrid earthworm was developed — New earthworm does not form mounds on lawns or golf course — Leaves its casting under the surface near the root zones — Characteristics of the new worm — Retaining all favorable characteristics of both compost and orchard worm — Has no unfortunate characteristics

Science has admittedly known and appreciated the work of the earthworm for well over half a century. Many farmers, orchardists and gardeners have realized that in soil in which earthworms lived, plant and vegetable life prospered.

There are scattered instances where farmers fertilized land with decayed animal matter hauled from stables. They attempted to transplant the compost-bred earthworms. These attempts have been recorded, but, no known sincere effort was ever made to discover why the earthworms perished when moved.

Consistent experiments and research work showed that earthworms are as much in need of the food on which they were raised as fish are in need of water. It was found that compost-bred earthworms demanded decayed animal matter; those raised in soil containing decayed vegetable matter demanded humus.

The author's first efforts to develop a satisfactory cross-bred earthworm were made in 1927. Selected specimens of earthworms found in various sections of the United States were studied, bred and interbred.

While engaged in landscape artistry, the author made observations of the brandling. It was shown that the brandling possessed many favorable qualities that could make crossbreeding very advantageous.

Chief among these favorable qualities was the fact that the brandling never deposited its excretions above the surface of the soil. This has two very important advantages.

First, no mounds are formed on the surface of the soil. Such little bumps cause lawns and golf courses to become uneven, sometimes unsightly, and, in the case of a golf course, ill-suited for the enthusiast of mountain billiards.

Second, all its castings are left under the surface of the soil near the root zones. The roots of plants and vegetables have easier access to the chemical and mineral elements pulverized by the earthworm's digestive tract.

Early experiments with the brandling showed that it appeared completely content in a tray, box or can. As long as it was well supplied with food, it was a prolific breeder.

Another characteristic of the brandling was its habit of living close to the surface of the soil, seldom going below six inches. Such a burrowing earthworm will cultivate the soil only around the upper roots of the plants and vegetables. While this may produce satisfactory results for some plant life, the author's desire was to develop an earthworm that would penetrate deeper into the soil.

Search for a promising earthworm to mate with the brandling produced no satisfactory results until a variety of orchard worm was found while matured trees were being transplanted.

This worm was large, and apparently spent much of its time deep in the ground, often down to ten and twelve feet.

A number of these worms were procured, carefully fed and studied. Observations showed that they burrowed as deep in the experimental trays, boxes and cans as they could get.

This type of orchard worm seemed to be an ideal medium for experimentation in the hope of producing a fertile cross between it and the brandling. Healthy specimens of both were selected.

These were placed in a special soil mixture, approximately one-third soil, one-third vegetable humus and one-third decayed animal matter. Such a composition of inorganic contains all (and theoretically more) of the vital elements necessary for plant life. These elements, however, are not always available to the roots of plants, as was explained in the introduction to this work.

Henry Drummond (1851-1897), an English philosopher and writer, pointed out that:

> "The inorganic or the mineral world is absolutely cut off from the plant or animal world... No change of substance, no modification of environment, no form of energy, no chemistry, no electricity, no evolution of any kind can ever endow a single atom of the mineral world with the attribute of life. Only by the dipping down into this dead world of some living form (Drummond obviously referred to the roots of plants, and we may safely add earthworms) can those dead atoms be gifted with the properties of vitality. Without this contact with life they remain fixed in the inorganic sphere forever."

Some form of life, either plant roots or earthworms, must bridge the gap between inorganic and organic (or living) matter, before the inorganic matter becomes available to plant life.

In the course of time, the worms having copulated, the egg capsules were extricated from the soil and placed in a separate container. When these hatched and grew to near-maturity, the weaker and less promising were culled out.

During the first six months, about a thousand cross-breeds which had been selected as breeders were mating and producing fertile eggs.

While this experiment seems to be the personification of simplicity, it should be realized that a full five years were consumed in these experiments. However, the results obtained in orchards, nurseries, gardens and poultry houses have proved that this quintet of years was worth every discouraging set-back. These set-backs were too numerous to be listed here. It is enough to say that there were times when Nature appeared to be stubbornly antagonizing all plans, figures and calculations.

I call this cross between the orchard and compost worm *Soilution*. Its chief features are:

1. A prolific breeder.

2. A free animal, no longer a slave to one environment.

3. Its castings never form objectionable mounds above the surface of the soil.

4. It is not an extensive traveler or migratory.

5. It makes exceptionally good fish bait, for it is lively and lives for many hours when impaled on a fish hook.

Lesson 4

Potential Markets for Earthworms

::

*Know your business — Unsound to enter any
business without knowledge of it — Fishermen
are possible customers for earthworm breeders
— Types of worm best suited for fishermen
— The orchardist needs the services of the
earthworm farmer — Orchardist is best
customer for earthworm culture bed — Farmers
are potential earthworm buyers — Poultrymen
can save money with earthworm culture beds
— One poultryman's opinion — The home
gardener is always interested in beautifying his
garden — Earthworms as garden beautifiers*

In discussing the potential markets for *Soilution* earthworms it is important to deal with facts and not become lost or confused in a maze of over-enthusiastic statements, over-zealous predictions. Neither should we imagine that overnight wealth awaits everyone who would enter this new development of a natural resource. It has been active on this planet long before man himself arrived.

That there is a wide and varied market for active, prolific earthworms is a fact too obvious to question. But these markets cannot be attained by a mere snap of the fingers. Financial security may also be assured, but certainly not by any magic power concealed in or about the culturing of worms for commercial use.

It is economically unsound for any individual to enter any type of business without at least a working knowledge of, or experience in, the business he or she selects. This economic rule, though undoubtedly it has its exceptions, is founded on sound logic and clear reasoning. It is the basic reason why beginning earthworm farmers should start on a small scale.

It should be kept in mind that the amateur earthworm farmer must thoroughly sell himself on the virtue of the earthworm. To accomplish this, he must study and understand the life and habits of his product. This may be accomplished by growing his own plants and vegetables as demonstrations with and without *Soilution* earthworm culture.

Once he has thoroughly sold himself, none can destroy the knowledge he has acquired nor the facts he has learned through personal experience, study and observation.

The tyro *Soilution* farmer must realize that it takes time for plants, shrubs and vegetables to show the benefits derived from the persistent and efficient work of the earthworm around the root zones.

Many forms of plant and vegetable life show a marked improvement in from 30 to 60 days after earthworms have been placed in the soil around their roots. In some instances, however, an entire growing season is required to prove the full merit of this type of culture.

And so, in justice to all persons contemplating entering this interesting and profitable line of work, it is hoped that readers of this book will not invest large sums of money in worm farms under the assumption that they will be financial lords overnight. To use a common expression, "it just isn't in the cards".

Assuming that the previous facts have been accepted, let us now turn our attention to the potential markets for selectively bred earthworms.

When the subject of earthworm farming is ushered into a conversation and discussed as a business, the ordinary person will usually recall the difficulty he experienced in finding worms in his garden the last time he planned a fishing trip.

Too many potential *Soilution* farmers imagine that every fisherman for miles around might be a good customer of one who could supply him with fat and active earthworms for bait. They fail to realize that many other potential earthworm farmers have thought the same way. The result is that competition becomes both keen and cutthroat during the annual fishing season. This results in a highly undesirable commercial and unsound economic condition.

Many years ago, the prevailing price of earthworm fish bait was one dollar for a six ounce can containing from 40 to 60 mature worms. The depression, plus absurd competition, brought the price tumbling down.

Competition was so keen in Denver, Colorado, that an earthworm price war resulted in six ounce cans being retailed for five cents.

The price of a can of earthworm bait in 1937 in California ranged from 30 to 50 cents for a can containing not less than fifty mature worms.

However, there is a pleasant and encouraging side to this type of unrestrained competition. Inefficient earthworm farmers are forced from the field, and men employing ethical business principles remain.

These are the men who thoroughly understand the remark of Phil D. Armour, of meat packing fame, who often said, "Any fool can compete, but it takes intelligence to organize and produce a better article."

When the smoke of the Denver price cutting war had cleared away, a lone worm farmer remained in the field. He had refused to reduce his prices, because he contended he had a better earthworm for fishermen.

Being a fisherman himself, the Denver man knew he could catch more fish with a small, active worm than he could with lifeless sections of large worms. This man's belief has been frequently substantiated by tests carried on by both amateur and expert fishermen, in soft and hard water, in lakes, rivers and gurgling streams.

The *Soilution* earthworm, properly fed and properly placed on the hook, will live and remain active for many hours. In various practical tests, *Soilution* has competed with other types of earthworms, as well as with amputated pieces of earthworms.

It is apparent that the entrance to this particular market for *Soilution* earthworms should be carefully planned and thoroughly examined before making a decision. This suggestion is especially sound if the prospective bait-worm farmer contemplates gambling on the necessary investment required.

The progressive poultryman and game bird producer are promising prospects for *Soilution* culture beds. Both, particularly the poultryman, are rapidly learning the value of properly fed earthworms as an aid to better poultry and better eggs.

An example of the interest shown in worm culture beds by modern progressive poultrymen is emphasized by the following, extracted from a personal letter to the author late in 1937:

> "Six years ago I knew very little about the poultry and fruit business. Today I own a flock of 1,500 splendid birds and have been exhibiting at many of the poultry shows throughout the West during the past three years, and I believe I have garnered my share of blue ribbons. Incidentally, my fruit tops the market and I give the lowly burrowing earthworm credit for much of my success.
>
> In educating the public in the value of selectively bred earthworms under control, you are doing a commendable and highly educational work—a work that should prove beneficial and profitable to all progressive people who depend on the soil for their living."

This poultryman had a *Soilution* culture bed for five years. His records show, and they are substantiated by similar tests made by other poultrymen, that a laying hen will consume from five to seven *Soilution* earthworms daily. This amount seems to be the limit of both her capacity and her appetite for them.

Similar tests on ducks produced startlingly different records. One small flock of carefully watched growing ducks consumed a gallon of earthworms daily and repeatedly quacked for more.

At this point the reader may ask why, if earthworms aid in producing better hens and therefore better eggs, poultrymen have not turned in greater numbers to the operation of earthworm culture beds.

This question deserves more than passing attention here. There are two very sound answers.

First, many individuals have entered the poultry business with the false assumption that all that was necessary to do was to use whatever feed the feed store operator sold or recommended. In the past, starting in the crazy twenties, hundreds of thousands of Easterners and Middle Westerners were encouraged to trek to California and enter the poultry business.

From 1920 to the year of the crash, a countless number of otherwise sane persons proved susceptible to the poultry raising bug in the southern portion of the Golden State. A vast majority of these individuals knew no more about poultry raising and breeding than an Amazonian native knows about a full dress suit.

The most many of these would-be poultry raisers contributed to the industry was a number of discouragingly black pages in their own personal book of experience. Many of them found that their dream of fortune became a nightmare of misfortune.

In a lesser degree, these conditions prevail in various sections of the United States and emphasize what has been said in the third paragraph of this chapter.

There is a second reason why so many poultrymen are laboring long hours and using feeding methods that should be classed as belonging to the horse and buggy age. It is the fact that many poultrymen, in many sections of the United States, are indebted to the feed man. These men dare not change their system of feeding their flocks for fear of reprisals from the feed man.

This is a discouraging situation, undoubtedly due to our nationally strained economic system. What hope there is for these men, what exit there is from their present position, are subjects more in keeping with a book on political philosophy or economic reform than for this volume.

The free-from-debt-to-the-feed-store poultryman is at liberty to purchase his poultry necessities, at liberty to operate his business, without being forced to abide by semi-dictatorial orders from outsiders. This man has left behind him the former methods of more or less haphazard feeding and has gone forward.

Through the assistance of the earthworm, the progressive poultry man can produce eggs for less than ten cents a dozen—a surprisingly low figure. In addition, he can increase the productive longevity of his birds and reduce the mortality rate of his pullets—a rate that has exceeded 50 percent in the state of California, according to reputable reports.

Since a single laying hen will consume about 2,000 breeding worms annually, the poultryman must operate a fairly large sized culture bed.

This is why the progressive poultryman may be considered by earthworm farmers as a very good potential customer.

The truck farmer, the suburbanite with a small vegetable garden, the nurseryman, the home gardener and the orchardist—all of these are potential buyers of properly bred, properly raised and properly fed earthworms.

Of this group, the one most in need of the earthworm as a natural cultivator and fertilizer is the orchardist. Then the rest in this order: the truck farmer, the small vegetable gardener, the nurseryman and the home gardener.

To state that an orchardist can reduce his overhead 50 percent by impregnating his soil with earthworms, may sound absurd. The following facts, carefully examined, checked and re-checked, should cause the most cynical person to realize that the immortal Charles Darwin did not exaggerate when he said that the earthworm is one of the world's most important animals.

Late in 1937, the following article appeared in the *Valley News*, Montrose, California:

"Near Redlands, California, is an orange grove that people come miles to observe. It demonstrates a unique natural method of orchard culture.

This 40 year old grove stands out among its neighbors in a way that even a layman can see. The foliage is thicker, a richer green, even at the top where others of its age show thin foliage and bare twigs. The trees are well filled with fruit and records show that they produce crops just as outstanding as their appearance. But the truly remarkable thing about this grove is the fact that these results are obtained with less labor, less water, and less fertilizer than is used by any of the neighbors.

The present owner took possession 17 years ago. Since that date, no plow, harrow or cultivator of any kind has been allowed in the grove. Weeds have been eliminated by hand labor. At first this caused extra expense; but since no weed is allowed to go to seed, a few hours labor once a month is now all that is needed.

The absence of mechanical cultivation is the first puzzle which this grove presents to horticulturists, for the necessity of soil conditioning has long been recognized. Actually this need has not been ignored here, but the owner depends, not on machinery, but on the world's finest and most efficient plow, the lowly earthworm. He has created conditions which are favorable to earthworms and in response they have multiplied until they are more numerous than in other groves. Their network of burrows has aerated the soil far more effectively and much deeper than mere surface cultivation could hope to do. At the same time, the feeder rootlets, which in an orange tree are very near the surface, are left undamaged, and therefore ready to absorb a maximum of food.

Even more puzzling to the orthodox grower is the fact that this grove thrives on less than 50 percent of the water required by others. The answer once more is explained by the burrowing habits of the earthworms. They prefer the cooler soil under the trees and dig most of their burrows there, with very few out in the sunny spots. During irrigation; a large proportion of the water enters the soil through these burrows, with the result that most of it goes under the trees where the roots can use it, while much less than usual is wasted out beyond the root zone.

But the fact about the grove which seems hardest of all to comprehend is its fine health in spite of what seems to be a very inadequate fertilization plan: a little synthetic nitrate occasionally, nothing else in 17 years. Once again the earthworms furnish the answer, this time by their digestive processes. Earthworms depend for food on dead organic matter, leaves, old roots, etc. Through digestion these substances are changed in character so that they are highly soluble and when ejected are immediately available as plant food. A close examination of litter under the trees reveals thousands of leaves which have been completely consumed except for a delicate skeleton composed of their veins. The worms have put this material back into the soil, for reuse by the trees. Without them, it would be a very long time before the same material would become available for plant food.

The earthworm's gizzard triturates large quantities of soil which the earthworm takes into its body for two purposes — one to make his burrow by eating his way in; the other to obtain from his food all the essential elements necessary to produce fertile eggs.

New surfaces are thus exposed to the dissolving action of the irrigating water, and plant food elements are released which would otherwise remain locked up inside the grains of soil. Couple this with the fact that earthworms work to a depth of 6 or 8 feet, constantly bringing new dirt from these levels to the surface, and it can easily be understood how trees can thrive for a long period without the addition of new feed elements to the soil.

Earthworms are nature's own means of soil building and conditioning. No orchard or garden can do its best without them. There are many kinds, some much more effective than others, and the study of their use and culture will repay anyone who grows fruit and flowers."

The article concerns fine, cultivated fruit trees, and the reader might consider them so developed that they respond easily and quickly to such experiments. The antecedents of these trees have been more or less pampered by orchardists for many generations. Man, in his desire to force them to bear more and more fruit, has grafted and pruned and fertilized and sprayed. The fact remains that they have not responded to earthworm culture more easily or rapidly than the woody perennial trees growing wild on mountains and in forests.

Seedling pine trees have been impregnated with *Soilution* for the Forestry Service in the Sierra Madre Forest Reserve. Each treated tree is clearly marked and identified. The pines treated have grown in two years to a height usually requiring five years to reach.

On March 23, 1937, a wet, heavy snow blanketed the region. Many four-year-old pines, whose soil was not treated with earthworms, were almost carried to the ground by the weight of the snow upon their branches.

On the same day, other pines which had been supplied with the elements Nature intended them to have, stood perfectly straight. These necessary elements were made available to the roots of the pines through the pulverizing action of the earthworms.

An ambitious earthworm farmer may very easily demonstrate the ability of the earthworm as a cultivator, triturator, chemist and distributor. He may do this on his own premises and for the interest of any visitors seeking to improve the quality and quantity of their gardens, orchards, farms or truck patches.

For any thoroughly interested individuals, the farmer can offer to place demonstrations on their properties. Then, they can see for themselves how the earthworm improves plants.

Controlling Production

There is a method of controlling egg production of *Soilution* earthworms which makes it possible to have a crop every month in the year. This information is given only to people who wish to go into the business commercially.

The Earthworm

Little brown earthworm under the sod,
A trusted worker in Nature's plan,
Fulfilling his destiny, obeying his God,
Living his life as a friend of man.

East, where the Nile flows down from the hills
Covering the sun-baked thirsty soil,
An important place in the scheme he fills,
Bringing new life by his humble toil.

West, where the pioneer follows the plow,
Turning the sod to the sun and rain,
This little brown brother is doing his bit
To nourish the roots of the growing grain.

North, where the snow lies still and white
And the blustering wind blows cold and chill,
When the spring thaw comes you will find him at work
And October's Harvest can go to the mill.

South, where cattle roam Argentine's plains,
And grass grows tall 'neath the summer sun,
This busy fellow is doing his best
To make "two blades grow in the place of one".

— *K.J. McCreedy*

Part 2

Introduction

::

Every living thing, be it a tender blade of grass or a giant oak, an infinitesimal germ or a human being, has one point in common with every other living thing—it must eat.

And, though eating is as commonplace to human beings as breathing, most of us dig our own premature graves with our teeth. If our teeth haven't given out before the grave is ready to receive the remains of a badly mismanaged and mistreated body.

In the light of present-day science and the ease with which knowledge is acquired, one is forced to stand agape at the general apathy the public maintains in matters pertaining to food.

We need not go beyond our own circles of friends and acquaintances to have this fact brought to our attention. Who among us doesn't know of an overweight woman who greedily devours every dish before her, especially those containing sweets and starches? Or a thin, sickly woman who pecks and nibbles at what is placed before her?

Overeating and undereating bring premature destruction to countless thousands of individuals annually. Out of these two, the undereater is a subject with whom we might well sympathize. Usually, this type of individual is putty in the hands of all kinds of food faddists, new idea dietitians and their ilk.

Someone long ago passed out the misinformation that the human body is a machine. Ever since that unfortunate moment, faddists and charlatans have hooked their financially profitable ideas to this false idea, and have "gone to town with it".

Fundamentally, the human body is no more like a machine than modern printing resembles the crudest of prehistoric methods of record keeping.

The human body, as Alexis Carrel clearly explains in his remarkable work *Man, the Unknown*, originates in a single cell and grows into a series of cells. These ultimately become the manifold unity of a living, breathing individual.

A machine is brought into being by an entirely opposite method. First, instead of one small unit—the single cell that is man's beginning—there are hundreds of small parts. When properly assembled and fitted into their designated positions, these make one complete unit, ready to function as its designer planned.

Feeding living organisms—plant or animal, poultry or man—should not be looked upon in the same light as one considers fueling one's automobile. In the latter case, gasoline, lubricating oil and water are required in all but a few air-cooled motors. The gasoline is the actual "food", with oil and water playing their roles of lubricating and cooling the mechanism so that the gasoline will generate power, or, to fit the analogy, life.

Living things demand more than one, two or three essentials to continue to live. The living organism is a highly complicated unit, with each component part requiring definite types of foods or fuel.

This volume is primarily designed to assist poultry raisers and breeders in developing a sound, more economical system of producing better eggs at a lower cost, thus, producing better, healthier poultry. It is also the hope of this volume to explain food values which may be applied by human beings to their advantage.

The writer does not accept the Freudian theory that each individual's life is controlled by his or her sex glands. He does accept the scientific and biological fact that within the sex glands of all living things are, by Nature, the chief chemical and mineral elements necessary to procreate.

And so, from this premise, I approach a subject previously neglected.

What is Food?

::

Various names for food — Alexis Carrel, Nobel Prize winner, quoted — Devitalized foods — White flour — Refined sugar — Pasteurized milk — We need a better Pure Foods and Drug Act — How children suffer from improper foods — Potency of procreative glands in animals and plants — How religion has bred ignorance on this subject — What are vital food energies? — Errors poultrymen make in feeding their flocks

For the purpose of blazing a straight trail to the goal of *Part 2*—better eggs and poultry at revolutionary low cost—it is helpful to consider food as nourishment for plant and animal life, including man. In a general sense, everything animals eat, and plants absorb through their roots, is food. In a more strict sense, food is any solid matter taken into the systems of plants or animals which serves to build up physical structure.

A food may be extremely pleasing to the eye, smell and taste, yet have no more food value than a toothpick. Much of this deception was generated in commercial and industrial interests who were selfishly seeking financial gain rather than preparing and marketing food stuffs with high nutritional content. While these commercial and industrial movements were heralded as being for the common good, the benefits derived by the public are, at best, debatable.

We will use these three staple foods—bread, sugar and milk—to bring light to this fact.

Alexis Carrel tells us in *Man, the Unknown*:

> "Our life is influenced in a large measure by commercial advertising. Such publicity is undertaken only in the interest of the advertisers and not of the consumers. For example, the public has been made to believe that white bread is better than brown. Then, flour has been bolted more and more thoroughly and thus deprived of its most useful components. Such treatment permits its preservation for longer periods and facilitates the making of bread. The millers and the bakers earn more money. The consumers eat an inferior product, believing it to be a superior one. And in the countries where bread is the principal food, the population degenerates. Enormous amounts of money are spent for publicity. As a result, large quantities of alimentary and pharmaceutical products, at the least useless, and often harmful, are thought to be necessary for civilized man. In this manner the greediness of individuals, sufficiently shrewd to create a popular demand for the goods that they have for sale, plays a leading part in the modern world."

Dr. Carrel's analysis of the promotion of white bread may also be applied to the popularity of refined sugar and pasteurized milk.

In the case of sugar, many of the most important elements have been refined out of the raw material, giving us a devitalized product. We know that refined sugar "looks better" on the table than raw sugar; that it is easier to shovel from the sugar bowl to the coffee cup and that it keeps indefinitely in our cupboard. These things we know, but advertising and publicity has led us to imagine refined sugar superior to raw sugar.

Pasteurized milk is the third staple which modern advertising has tricked us into believing is superior to raw milk. Regardless of the fact that pasteurization completely destroys many of the vitally important elements in milk necessary to good health.

Summed up, white flour, refined sugar and pasteurized milk are counterfeit foods passed off on a gullible and apathetic public.

A large percentage of our present day ills may rightly be traced to deficiencies in our food. Many of these deficiencies are traceable to the high-speed and high-production systems employed in modern plants concerned with the business of making, packing and canning food stuffs.

This is civilization. Perhaps the Oriental Sage was not far from the truth when he defined civilization as a deterrent to progress.

Every civilization has within itself a suicide germ. This germ is fed by collective and individual greed. It destroyed the Greek, the Chinese and the Roman civilizations, and, as we have seen in the introduction to this volume, it is gaining potency in America through the medium of monopolies. All about us on this whirling sphere, national civilizations are cracking and crumbling. American civilization is cracking—every reasoning person is cognizant of this. Will it crumble? That is a question only time will decide.

No race, no nation whose members are both physically and mentally deficient or deteriorating can stem the encroachment of racial or national destruction. And no race or nation can expect its members to increase their physical and mental development if it persists in permitting misinformation about food deficiencies to continue.

"But we have a Pure Food and Drug Act on our statute books in Washington," you say. Yes, there is such an Act, but it is as useless as a gunless submarine on the Mojave Desert. If this act were rewritten, sincerely rewritten for the benefit of you and me and our children and their children's children, every sack of flour, every bag of refined sugar and every bottle of pasteurized milk would something like this written on it, in large letters:

THIS IS A DEVITALIZED PRODUCT.
IT HAS VERY LITTLE FOOD VALUE.

But the mere mention of a truly sincere pure food and drug act sends shivers up and down the spines of the financially powerful milling, refining and kindred industries. Those whose leaders commit the sin of omission by refraining from telling their customers the true facts about devitalized foods.

Dr. Carrel merely brushed the surface when he wrote that

> "...in the countries where bread is the principle food,
> the population degenerates."

The unnatural conditions that follow in the wake of a continued diet of devitalized foods are destined to take their toll in weaker physiques, duller mentalities and in human lives.

The current generation of children is probably suffering more from food deficiencies than the preceding generation. This is partly due to the fact that a large majority of American parents have been forced, through economic conditions, to buy cheap food, or food of which they receive quantity rather than quality.

Another important factor, perhaps the chief factor, is radio advertising. The popularity of radio has made of this medium an ideal outlet for the fancy and romantic, albeit questionable, phrases of advertising copy writers:

Eat Whitey's Wheat Wafers and become a football star.

Breakfast on Betty's Baked Barley Blocks and win a husband.

Drink more Pasteurized milk and lick your weights in wildcats.

*Give the kiddies Carter's Coddled Candy Cakes
and watch how quickly they grow.*

... and so on, *ad infinitum*.

Radio advertising is the height of psychological suggestion. Few are the radio fans capable of entering a drug or grocery store without leaving with a bag, package or bottle of a product their favorite radio performer says is "tops."

In the relatively pathological scramble for more business, manufacturers of food stuffs are permitted by an apparently disinterested government to flood the nation with devitalized foods for man and beast. Not a few of which are as valueless for nourishment as a rubber band.

This lack of necessary food elements in a growing child's diet results in the child actually gorging itself in an unconscious effort to obtain sufficient and proper nourishment. The stomach of such a child begins to distend, it is forced to distend in order to accommodate the unnatural and unreasonable amount of food it receives. In time, the child is never satisfied unless it has packed its unnaturally dilated alimentary canal until it feels, and actually is, full.

Though it is not generally known, the average American eats nearly five times more bulk than he needs. If this absurd condition continues, physiology text books in the years to come will refer to Americans as a mongrel breed of human beings noted for their Gargantuan stomachs. Certainly, this is not a complimentary prediction. Yet even today there are many stomachs distended to ten times their necessary size.

The consistent eating of dead or devitalized foods is quite probably responsible for the increasing number of sterile men and women in America under thirty years of age.

The procreative glands, being composed of the richest elements that have been transformed by the various organs of the body, cannot be supplied with those elements if they are not in the food consumed.

Continued dieting by motion picture actresses to remain slender to meet the exaggeration of size produced by the cinematic camera has not only injured their general health, but probably their procreative glands as well. When the body cannot receive sufficient elements from the food intake, it automatically turns to the procreative glands to supply the deficiency. This reservoir is eventually drained, resulting in a sterility that, more often than not, becomes chronic.

However, this reference does not mean that all normally-sized or obese women (or men) are fertile. Undue fat on the human frame does not denote health; usually it signifies the opposite. Fat-producing foods contain very insignificant amounts of the elements demanded by the procreative glands.

Except in instances where a physical abnormality is responsible, eating too much of the improper mixture of food may be rightly blamed for the supercargo of avoirdupois that is being hauled around by obese men and women.

However, human beings are not the only animals that gorge themselves. Many plants and domesticated animals glut. Not because it is their nature to do so, but because their diet is deficient in one or more elements necessary for good health.

All of this may appear to be a roundabout approach to our subject—poultry. Nevertheless, we shall presently observe that what has been said about food for human beings, aptly applies to food for all forms of life.

Let us now consider the difference between organic or live food and devitalized or dead (inorganic) food.

It is at this point that the classical bull is taken by the horns and a subject discussed that has been shunned by writers on food and food values—the unequaled nutritional value of the procreative germ as food.

Our subject must be approached gently, for religious thinking in the past warns us that sex, in any form, is more or less taboo.

But no matter what one's religious theories may be, the fact should be accepted that life begins with sex. All living things owe their humble origin to sex, regardless of the method Nature employs in planting the fertile, procreative germ.

This germ is infinitesimal. Yet, Nature has combined all—not one, or a few, or an incomplete group—but *all* the vital necessities which, when blended with its direct opposite, produces a human being, an ant, an elephant, an orchid.

The vital qualities that are in the germ-plasm are what our bodies need. We receive one type of them when we eat wheat that has not been devitalized. Such a food is a "live" or "organic" food. When we eat bolted wheat, most, if not all, of these vital qualities are missing. Bolting has destroyed them. Such food is "dead," or "devitalized." Its nutritional value is practically nil.

Fruits, berries, nuts and all grains, if not devitalized by

dehydration or cooking, are the quintessence of the richest elements from the mother plant, tree or bush.

However, these germ cells are not rich—some are even sterile—if the plants, like human beings, do not receive their necessary elements from their food, the soil.

The average person has no qualms about eating the germ cells of fruits, berries, nuts, grains and vegetables. Nor does he object to eating eggs, which, if fertile, contain procreative germs.

The genitals of domesticated male animals—those whose flesh we approve as edible—are both palatable and nourishing. Most persons balk at the mere thought of such a dish—a delicacy when properly prepared. They will eat a fertile egg without a quibble, but make a wry face if the above dish—colloquially known as 'lamb, sheep, pork fry," or "Rocky Mountain oyster"—is suggested.

Carnivorous animals in their wild state will gorge themselves, not on the flesh of their kill, but upon the vital organs. Instinct instructs them that they will get the most nourishing elements from these organs for their bodies.

Rodents in a wild state are not interested in eating the lifeless leaves of plants. They live almost exclusively on grains, kernels, berries and seeds.

Are we to assume from these natural facts that wild beasts and rodents have "more sense" than we? Certainly, if one considers present-day eating, we would not be far wrong to admit their superiority in the matter of eating what is nourishing and what is not nourishing food.

Poultrymen are committing a grievous error in feeding their flocks tons of devitalized foods annually. Within their grasp is enough vital food to bring them greater profits, more productive hens and more fertile eggs.

How this germ-food for poultry and game birds is acquired is the subject of the next chapter.

The Germ Life and Better Poultry

###

This chapter deals with the second forward step toward better poultry through natural foods. We shall move along lines similar to those discussed in the preceding chapter.

This method of approach may appear unnecessary, but the writer has experience in these matters. There will be a short stop here to lay a foundation so the reader may have a better understanding of the method of producing healthier and more prolific poultry.

As we have already seen, the procreative cells of all animals and plants are extraordinarily rich in vital elements. Many other parts of edible animals and plants may be rich in food elements, but this richness is not necessarily native to the animal or plant. It is an acquired richness and comes from the food the animal or plant subsists upon during its life.

Here is an explanation of what is meant by "acquired richness":

Picture two chickens. One is penned up in an ordinary yard. The soil is dry and dusty. The food fed this chicken is one of any number of commercialized poultry feeds, most of which are devitalized. The other chicken is turned loose near the dairy. All day it scratches and pecks in the decomposed animal matter piled nearby. The owner of the second hen need spends no money for commercial feed.

The first chicken is suffering from food deficiency, whether its owner realizes it or not. The second chicken is enjoying a full life, replete with a balanced, natural ration of food from which its entire physical system draws potent nourishment.

How do we account for this? Simply by realizing the fact that the chicken pecking in the manure pile has been eating *live food*, the other, *dead food*.

That one of these foods is dead, the other alive, is easily demonstrated by taking a shovelful of each and planting it. The shovelful from the manure pile will, in due time, send up sprouts of grain, depending upon that eaten by the animal that dropped it. Not all live grain germs are destroyed by an animal's digestive system—not even in the dual systems of ruminants.

In comparison, the shovelful of dry soil, having no form of plant germ life, returns nothing for the effort expended in planting it.

The foregoing example brings us again to organic (living) and inorganic (lifeless) matter. Both the dry, dusty soil in the chicken yard and the pile of manure back of the dairy barn are scientifically classified as inorganic matter.

However, there is considerable life in manure. It contains millions and millions of bacteria, which, as everyone knows, are infinitesimal organisms. Bacteria are also known as microbes and germs.

At this point, the reader should accept the fact that there are beneficial as well as dangerous bacteria. Some of these, according to the germ theorists of the medical fraternity, are disease germs and will, if left alone, produce illness and death.

Probably man's greatest handicap in this so-called enlightened age is not so much the general ignorance that is prevalent regarding Nature and her methods as it is the misinformation and misapplication of known facts. These are distorted by some "authorities" for their personal and collective financial gain.

That certain bacteria are destructive to animal and plant tissues is a fact—a natural fact. But Nature, in her own definitely evolutionary and omniscient way, has developed other bacteria to defeat the dangerous inroads of destructive bacteria. This is one form of the law of preservation. Were Nature to develop one bacterium, one small insignificantly-sized germ, without also developing an enemy to keep it under control, entire species of animals and plants would pass from this planet.

Once destruction appears in living matter, if Nature cannot combat it, she appears to hasten its ruin. The result is ultimate death. And even here, she does not stem her onward march to destruction. Decomposition is rapid in most instances of dead, once organic matter. Nature is anxious to prepare that matter for consumption by other living bodies.

The late Ambrose Bierce, the immortal San Franciscan, described 'edible' in his *Cynic's Dictionary* as

> "good to eat, and wholesome to digest, as a worm to
> a toad, a toad to a snake, a snake to a pig, a pig to a
> man, and a man to a worm."

(It is worth a parenthetical note here to say that only the lifeless remains of once living matter decays and decomposes rapidly. Inorganic matter, like rocks, for example, do decay and ultimately decompose, but the process is slow and drawn-out, lasting thousands of years. The reason for this is explained when we realize that Nature can put such inorganic matter to little use as food for her various species.)

Bacteria, good and bad, have their place and their function in the laws governing all living things.

These minute creatures are everywhere—in the air we breathe, the water we drink and the soil we plow. Without them, you and I would not be here. We—all living animals—feed, to a greater or lesser degree, on plants. Plants in their turn, feed upon water, carbon dioxide and the nitrogens and other salts.

We know where plants receive their needed water and carbon dioxide. We should know that they receive their necessary nitrogenous salts from but one source—bacteria—and that bacteria, in their turn, receive their necessary nourishment from a few minerals.

In short, bacteria are the beginning of animation, of life. Beyond bacteria is inanimate matter, lifeless rocks and colloidal substances.

So we see that the chicken pecking in the manure pile has a veritable storehouse of living food, countless trillions of minute bacteria, rich in the necessary food values its feathered body demands.

In rich soil, too, soil moistened adequately with water, countless other trillions of bacteria exist. The eminent naturalist, David Starr Jordan, has pointed out that he has found over four million bacteria in one gram of such soil.

Let us now return to the procreative germ, the germ of life produced in healthy, mature animals and plants. This germ is the rudimentary element, the primary source of everything that lives, the earliest stage of an organism, the cause, origin, principle and prime mover of all life.

Though science has discovered many interesting and astonishing things about germs—both procreative and bacterial—there still remains much to be learned about these minute organisms. Many of these are so infinitesimal that they are barely visible under a strong, microscopic lens.

Until the World War period, the maggots or larvae of flies were considered as filthy and disease-bearing as any creature known to man.

It was the late William Shakespeare Baer, an American M.D. working with the French forces, who first came upon the bacteria-destroying ability of fly maggots.

Two *poilus* were brought to the hospital where Dr. Baer was stationed. They had lain for a week behind bushes, minus food and water. Each had a thigh horribly smashed, and the shattered bones protruded through the skins, and the wounds extended into each abdomen. According to all medical science of the day, both men should have been dead. Yet they were alive, conscious and hungry.

When Dr. Baer examined these men, he found their wounds a teeming mass of fly maggots. When he cleaned them away, the flesh and bones were bright and clear and healthy. There was not the slightest trace of a disease-bearing germ anywhere in the wounds. The fly maggots had destroyed them.

With the passing of years, Dr. Baer experimented with fly maggots. He worked on the theory that they were more efficient workmen than the finest surgeon with all his modern sanitary methods, sterilized instruments, operations and amputations.

Naturally, such a theory did not set well with certain members of his profession, but Dr. Baer didn't care. He thought of the two French soldiers, of how they lived when they should have been dead, and of the hundreds of children suffering from *osteomyelitis*—a pernicious rotting of bone marrow and bone structure.

In the Children's Hospital in Baltimore, of which he was the head, Dr. Baer bred fly maggots and placed them in the openings he made in the afflicted flesh. Ignoring objections made by the medical community, he carried on his work. He fought strenuously against those who had an aversion to anything new or revolutionary. Things that should be common knowledge.

And Dr. Baer cured those children—or, rather, the fly maggots cured them!

Though the story of Dr. Baer's work and struggle deserves considerable space, there is, unfortunately, no room for it here. It is enough to conclude that fly maggots are used extensively today by the medical community as destroyers of many disease germs. All of this proves the point desired—that many destructive germs can be destroyed if other friendly germs are given a chance to work for us.

For centuries, the medical profession has been teaching the general public to scowl at the mention of the word "maggot." And so, the average person has generated both an aversion and a hate for the word. One may actually question if its use in polite society is permissible. To the average mind, "maggot" is synonymous with filth, offal, danger and disease.

And now the medical profession finds itself in the ludicrous position of expounding something it has up to now warned against.

Appreciating the general dislike for the word "maggot," it will appear from now on in this work under its other name: "larva."

Thousands of germs never pass the germ stage. That is to say, certain germs live and die as germs. Others go through varied forms of evolution. All germs begin life as single cells. Those that are destined by Nature to evolve into a higher plane, join another germ.

Nature employs a variety of systems to perform the task of impregnation. In higher animals, including man, the single male and female germs blend and produce the embryo. The evolution of the single cell germ is made in one remarkable step.

In certain insects, the fleshfly, for example, the change from the procreative germ to the living, ultimate fly is performed by a circuitous route.

The fleshfly, of which the bluebottle is probably the best known, lays her eggs, many hundreds of them, in decayed or decaying animal flesh. In from two to three days, these eggs hatch into small, white worms or larvae. In time, a sort of shell or cocoon forms around the worm. It enters a dormant period, during which the worm is being slowly transformed into a fly. Upon the completion of the transformation, a full-grown bluebottle appears.

It is the intermediate stage that spans the egg to the fly with which we are especially concerned, and in a later chapter it will be referred to again at some length.

Thought, immaterial though it may be, is the matrix that shapes the issues of life. The mind has been active in all fields during this fruitful century, but it is to science we must look for the thoughts that have shaped all thinking.

Economical Poultry Housing

::

*Poultry business in California — Brief
background of poultry colonies — Promises
made by promoters — Poultry housing costs
— Buildings designed by "experts" — Cost
exorbitant — Modern hennery simple to build
— Cuts housing cost down from one-tenth to
one-thirtieth of general cost — Better chance for
profits when poultry housing investment is low*

In California, particularly Southern California (and in other sections of the United States to a lesser degree), there are acres and acres of empty and dilapidated poultry houses. These slowly disintegrating buildings are the ghosts of the dreams of thousands who entered the poultry business intent on making a fortune "selling eggs." These crumbling, grim reminders of individual tragedies stand today as potent examples of what usually results from get-rich-quick schemes.

In addition to these ghosts of failed dreams, California is dotted with the decaying remains of poultry colonies. Highly-touted, co-operative organizations with the magnetic slogan "One for All and All for One." Usually this plagiarism from Dumas' *The Three Musketeers* proved beneficial to but one, and that one was the promoter. Gullible prospects invested in these colonies for a few dollars down, the balance to be paid from the profits made from the poultry.

Great, indeed, have been the glowing promises made to potential poultry farmers, especially in Southern California. The ghost towns of the gold regions of the state are insignificant in number when compared with the ghost poultry farms in Los Angeles County alone.

It appears safe to state that a vast percentage of those failures were due to the unnecessarily high cost of housing the poultry as well as the high cost of feeding.

This chapter deals with the problem of housing the poultry economically. In the following chapter, the problem of feeding poultry economically will be discussed.

According to so-called scientific methods, at this point in time, housing poultry runs from one to three dollars per bird. These henneries are frequently designed by "authorities," but when these "experts" are investigated, one usually discovers that they are men who have never raised a chick. While these houses appear attractively practical, they lack much that a poultryman needs if he expects to make a profit on his investment.

The cost of these poultry buildings range from one thousand to three thousand dollars and are intended to house up to a few thousand birds. Considering the maximum returns from eggs and meat birds, such an investment is definitely out of proportion.

All of this probably sounds fantastic to the average active poultryman. Yet, if he is honest with himself, he will agree that simple mathematics prove the fallacy of such an investment for poultry housing.

I am personally acquainted with many active poultrymen. They are hardworking, conscientious men, laboring an average of twelve full hours every day of the year, including Sundays and holidays.

Some of these men have sound, healthy flocks. Their business should show a net profit of from one to two dollars per bird per year. Many of them keep no books and cannot come within fifty percent of accurately estimating what it costs them to produce one dozen eggs.

I have discussed poultry housing costs with the most successful of these men. None of them have taken issue with me when I've said that any poultryman whose housing costs run to one dollar a bird—which is probably an average cost—must face the undeniable fact that, from eggs alone, a hen will spend her entire life paying for her housing.

It is my contention, that any poultryman whose housing cost runs above fifty cents per bird is destined to ultimate failure unless he is financially able to forget the interest due him on his original investment.

But, you will say, you have been told that in a couple of years you will have paid off the original investment in the building or buildings out of the profits of the business.

Let us figure this profit.

Out of the revenue brought in by the hens, after deducting for the mortality, feed, wages, water and whatnot, one must consider the original cost of the flock, taxes, insurance and interest on the money invested in the ground space used. Whatever is left may then be applied on the original housing cost.

One can readily understand how long it would take a hen to pay her housing, assuming that the net profit of the hen (that is, the amount over and above the expense of her general upkeep) should be from one to two dollars per year. It is easily understood that the struggle to meet this indebtedness is usually extended over a period of several years—if the poultryman lasts that long.

Let us concern ourselves now with the writer's idea of a wholly practical and extremely economical hennery.

Let us assume that we have approximately 400 square feet of land available for use for a hen house. On this we shall construct a hennery for the housing of 100 fowls.

In shape, our hen house may be square, rectangular, round, hexagon or any shape suitable for the space to be used.

The construction of the hennery is simple. It may be built where it is to be placed, or in sections in the workshop. The material used is reasonably priced—and new material need not be used if other suitable material is at hand. This consists of lath—or batting if stronger construction is required—one by two inch boards and wider boards for the bases. Half inch pipe is sufficient to supply water; some wire netting and a small amount of hardware complete the building requirements.

Including labor costs, whether of your own or those of a handyman, the total cost of such a hennery should be in the neighborhood of fifteen dollars—or fifteen cents per hen! With prices differing in various sections of the United States for chickens, and with certain breeds running higher than others, the writer at no time attempts to include an estimate of the cost of the flock.

This fifteen cents per bird is a surprising figure, as any active poultryman will agree. It is from one-tenth to one-thirtieth of accepted poultry-housing figures—and it should take the hen from one-tenth to one-thirtieth as long to pay for it!

Some years ago, the writer constructed a hennery to prove to skeptics that his theory regarding hen houses and poultry feeding was practical. No effort was made, nor was it intended, to make this hen house anything but an experimental and demonstrational enclosure. Nonetheless, it definitely and conclusively proved its worth and admirably demonstrated the fact that better poultry can be successfully raised in compact confinement.

The reader should not jump at the conclusion that this type of hen house requires the elimination of a "scratching shed" or "run yard" beyond the confines of the hennery proper. If space is available, it may be used, especially if the poultryman insists his hens need more room for exercise.

What is inside this small, compact, yet wholly practical hennery? Having a clearance from ground to latticed roof of but two feet, there cannot be much room for anything, you may contend.

But there is all the room needed when the hens are scientifically fed.

I can best describe the "furnishings" of this hennery by taking the reader into the next chapter. Here, the interior of the hen house is tied in tightly with the feeding of the poultry what I am pleased to call 'Intensive Range'.

The Interior of the Economical Hennery

::

Popular poultry feeding systems will not work with Intensive Range — Why poultry is destructive to gardens — Poultrymen should take advantage of this knowledge — Fowls should not be forced to glut to receive non-deficient foods — Two methods of feeding Intensive Range — Building the economical hennery — Where earthworms come into the picture — How to make an earthworm pit — Henhouses for large poultry flocks

In the preceding chapter, the writer maintained that it is possible to construct a hennery to accommodate 100 hens for the small sum of fifteen dollars. A more definite explanation of its construction was omitted because there are a few highly important features necessary to meet the requirements of a system of feeding which must be explained before actual construction of the henhouse can be described.

Any hennery constructed as described in the previous chapter, no matter how masterfully built, is not worth the space it covers if the poultryman does not follow the system of economical poultry feeding explained in this work.

To attempt to feed hens under any other of the popular systems in such a henhouse would spell ruin for the poultryman.

The writer's sole aim here is to point out to the reader the road to economical poultry raising by feeding Intensive Range. Proper housing facilities for this system of feeding are necessary.

We must not lose sight of the fact that devitalized foods, for man, beast or plant, cannot furnish proper nourishment. Such food is responsible for the unnatural overloading of the digestive organs. This point is emphasized here because we are about to describe the natural food for poultry, one that has been distressingly pushed aside—live food!

The poultryman does not have to be told that poultry is destructive to gardens. The reason for this is that the fowls prefer sprouted seeds, insects, larvae and earthworms to all other varieties of food.

Admitting this fact, why do poultrymen continue to feed their flocks dead or devitalized food?

This is a question over which I have pondered for forty odd years. Why? Why? Why?

The term "Intensive Range" was so selected by the writer because the food provided contains, in a small compactness, all the food elements that game birds and poultry can find on the range.

Intensive Range provides fowls with the necessary 18 percent of albuminoids, 7 of fats and 75 of carbohydrates, which are approximately the percentages Nature intended. Fowls fed on Intensive Range will not have to gorge themselves in an attempt to obtain these essential food elements.

But before I yield to the temptation to wander too far, let us return to the construction of the interior of our hennery.

There are two methods of feeding hens Intensive Range, each economically productive. One of these is by means of "trays." The other, by utilizing the soil of the floor of the henhouse.

The first system is best to use for smaller flocks of poultry and where the ground space is limited.

The second method is more suitable for large flocks, ranging upward of a few hundred.

We shall begin with the first or "tray" method.

If the poultryman with a small flock of birds constructs the type of hennery discussed in the preceding lesson, he should follow these general instructions:

Regardless of the shape it should be lathed on the sides and top with a space between the laths wide enough for the hens to stick their heads out.

Adequate roosts should be built halfway between the floor and the latticed roof, directly over a pit three feet deep which has previously been excavated, then filled with about one-half manure and one-half peat moss (wood shaving or sawdust, if moss is not available). It is into this pit that we put our friends, the earthworms.

In the writer's experimental hennery, fifty thousand earthworms were originally placed in the pit. This may seem a Gargantuan amount. Earthworms are prolific breeders, and when properly raised will multiply at an astonishing rate.

It has been demonstrated that this large number of active earthworms are necessary for the above-described pit. A lesser number cannot perform the expected work efficiently.

Once the pit is prepared, it should be covered with ordinary chicken wire, raised sufficiently to defeat any attempt of the chickens getting into it. Wire should also be placed upright around the edges of the pit to prevent the chickens from getting into the area from the sides.

The principle of the earthworm pit in the "tray" method of feeding is twofold.

First, the worms dispose of the hens' droppings, thus eliminating the unpleasant odor that is common to all ordinarily-operated poultry yards, and eliminates the disagreeable task of cleaning the henhouse.

Second, as the worms multiply, they burrow beyond the limits of the wired pit and thoroughly impregnate the ground. Once in this outer area, the scratching hens will unearth and eat them.

Repeated studies and observations lead us to believe that hens will not eat more than seven to nine earthworms a day, provided they are fed vitalized food such as that contained in Intensive Range.

The original filling of the earthworm pit is all that is needed. A hen's average droppings approximate seventy-five pounds annually. This amount of fertilizer (from one hundred hens) assures the replenishment of the pit permanently.

The water supply should be planned so that the overflow from the hens' drinking fountain will find its way into the pit which, at all times, should be kept moist to a degree bordering on sogginess.

Nests should be conveniently placed, both for the hens and the individual who collects the eggs.

The poultryman should also prepare an earthworm master pit, or culture bed, beyond the confines of his new hennery. This, for the sole purpose of furnishing castings for the "trays" for earthworm castings make the finest rich soil. It is in this soil that Intensive Range is prepared.

Intensive Range for fair-sized flocks is best prepared in wooden boxes. For small flocks, one hundred or less birds, I have found discarded wash basins, many of which may be purchased at secondhand stores for a few pennies, ideally suited for this work. However, almost any type of container may be used with more or less satisfactory results.

Intensive Range "trays" are filled with earthworm castings, peat moss (or substitute) and grain. (See *Lesson 5.*) The "trays" are then set aside and left until the grain sprouts. The contents are then dumped out and placed outside the pen, but within reach of the fowls through the slats.

For the second method of feeding Intensive Range—which is more suitable for large flocks—we must build a somewhat different hennery.

The same general principles are followed as in the first hennery for the first method of feeding with provision added for sprouting the grain in the scratching area of the pens.

Each hennery, therefore, must be divided into two compartments or pens, to be occupied alternately.

My personal opinion in regard to large flocks of poultry is that henhouses, similar to those already described, be constructed with accommodations for fifteen to twenty-five birds.

These buildings should be about 8 feet long, 8 feet wide and 2 feet high.

Earthworm pits are provided as in the first method, to take care of the hen's droppings and propagate the worms, some of which escape into the outer scratching area. The hens occupy one side of this enclosure while the grain, which has been sown in the other side, is sprouting in the scratching area. This area is now rich in earthworm castings from the worms that have made their escape from the pits.

The grain sprouts in seven to ten days, at which time the hens are moved back into this pen and the same procedure is carried out in the other compartment.

Assuming that I have covered the construction of these particular types of henhouses as well as this volume permits, the reader is now ready to learn more about Intensive Range.

Intensive Range

■■

Intensive Range — Its development through the years — More about devitalized foods — What Intensive Range will do for poultrymen — How to prepare it — Making the trays — Do not overfeed poultry — Fat hens are poor layers — Intensive Range is a balanced diet, replete with all of Nature's elements necessary for healthy birds — How to sprout grain

I approach this chapter as I imagine an actor would when he brings his leading character to the climax. This chapter is, indeed, a climax, not only of this volume, but of many years of experimentation. I have labored through many an anxious month awaiting the outcome of one or more experiments. And, if they failed, which many of them did, I hitched up my belt and started the nerve-racking tasks all over again.

In retrospect of the twenty years of experimental work with Intensive Range, it seems like only yesterday when I first attempted preparing a natural food for poultry in compact containers. Through the years, I have changed plans and experimented and changed them again, until today. I wholeheartedly believe I have succeeded in developing a natural food for poultry that will stand the severest scrutiny.

I would probably be charged with sophistry were I to promote Intensive Range as a perfect poultry food. But I fear no such charge when I state that Intensive Range has more of the qualities that constitute perfection than any other poultry food.

It should be noted that I do not refer to Intensive Range as a "formula." On a yearly basis, farm and agricultural journals predict a millennium (that never arrives) where there will arise a new poultry feed formula.

To the active poultrymen, these formulae are old, old stories. As yet, no genuine revolutionary commercial poultry feed has been brought to my attention, though steps in the right direction were taken when dehydrated seaweed (kelp) was added to poultry feed.

I say "steps in the right direction," because seaweed contains thirty-five of the forty-eight known elements found in the sea. As a food adjunct for man, beast or fowl, I cannot praise this marine plant enough.

Spasmodically exploited and usually short lived formulae merely add confusion to an already distressingly confused business. This confusion is traceable, I believe, to the various claims made by many commercial poultry feed manufacturers and dealers, wholesale and retail.

A large majority of these commercial feeds, many of them devitalized, are far removed from what is naturally the best regimen for poultry. These feeds are, in my opinion, driving the domesticated fowl farther and farther from their true nature.

In spite of the large part domesticated poultry plays in man's diet of the present day, these fowls and their eggs are actually late additions in his cuisine.

All of the represented types of poultry—Sebright Bantams. Black-Breasted Red Games, Dark Garnish or Indian Games, Japanese Bantams, Silver-Spangled Hamburgs, Pit-Games, La Fleche, White Leghorns. Black Langshans, Buff Cochins, Partridge-Cochins, Light Brahmas, Mottled Javas, Barred Plymouth Rocks, Silver-Gray Dorkings, Houdans, White-Faced Black Spanish, Black Monircas and White-Crested Black Polish—have come to us, through various degrees of breeding and crossbreeding, by a circuitous route from India.

To citizens of Burma probably goes the credit for being the first to domesticate fowls, perhaps about 2000 B.C. Later (about 1500 B.C.), domesticated fowls appeared in China. From there, they were introduced to Europe, and then to the Western Hemisphere.

No mention of fowls or poultry is made in the Old Testament, though eggs are referred to four times, in Jeremiah, 7:11, (about 600 B.C.); Isaiah 10:14. (about 710 B.C.); Deuteronomy 20:11-6 (about 1450 B.C.) and Job 6:6 (about 1520 B.C.). The New Testament refers twice to hens and thrice to cocks. It is also interesting to note that the early Britons would not eat domesticated fowls, tabooing them as they did the hare and the goose.

Until about half a century ago, most domesticated birds in the United States were raised on feed that contained much of the necessary food elements their feathered bodies required. Then, with the coming of Big Business, and, later, exploitation of poultry as a get-rich-quick enterprise, poultry began to be fed more and more on commercial feed.

For the most part, poultrymen of the last two or three decades seemingly have forgotten how their fathers and grandfathers raised chickens. The modern poultryman appears completely satisfied with the knowledge he receives from advertisements that blandly promise—too often in ambiguous phrases!—unbelievably wonderful results from the advertisers' feeds.

I have talked to many poultrymen who consider themselves modern. Many of these men look upon the manure pile as an abomination. In my heart, I actually pity believers in this form of modernism, just as I pity the individual who knows the earth is flat. Both individuals are in a mental mire from which no man can extricate them. Their only escape is through the proper use of their own reasoning faculties.

In offering Intensive Range to the poultry world, I am convinced that it will be received by two general types of persons.

On the one hand will be the poultryman who realizes the value of live food for his flock and will welcome the information I have to give. On the other hand will be the Sir Oracles who boast the virtues of their own infallible methods of feeding poultry regardless of the high mortality rate of their flocks. Their cost of producing eggs is not below that considered an average minimum and their feed bills run unreasonably high.

For the latter type of person, I paraphrase Emerson:

> "Condemnation before investigation is a barrier that will hold any mind in ignorance."

Conscientious preparation and feeding of Intensive Range will produce benefits all poultrymen seek, but few find. Each of these benefits, which I emphasize by numerical classification, is an imperative that all poultrymen covet.

Intensive Range will—

1) Reduce the feed bill about 50 percent.

2) Reduce the cost of egg production about 50 percent.

3) Reduce poultry mortality rate appreciably.

4) Reduce anatomical disorders in poultry.

5) Reduce poultryman's labor appreciably.

6) Increase the average longevity of fowls about 50 percent.

7) Produce more fertile eggs.

8) Produce better meat birds.

9) Eliminate "bare backs," or nearly featherless birds.

10) Eliminate henhouse odors and cleaning.

11) Eliminate the feeding of meat.

12) Supply the fowls with pepsin. (See *Lesson 6*).

These benefits are accomplished because Intensive Range is based on natural laws regarding food energy. Fundamentally, these laws evolve from the basic principle that all animal life is dependent, directly or indirectly, on vegetable life.

Ages before being domesticated, fowls lived on insects, insect larvae and the live seed germs of plants. If fowls permitted to return to their native state, they would soon seek out this kind of regimen. In time, they would probably redevelop their wing spread and power to take once again to the air, as their distant antecedents did.

Why, then, should we not feed these fowls what is naturally their proper food?

If the reader agrees with me that chickens should be thus fed, he will readily see how and why Intensive Range provides a natural food supply for them.

For small, fair sized flocks of poultry, Intensive Range is prepared, as previously noted, in trays.

The equipment and material needed consists of—

a) Earthworm castings—from the master pit.

b) Earthworms—not more than twelve.

c) Peat moss—or substitute.

d) Grain—barley, wheat or oats.

e) Fish Gills—or other fish, meat or fowl offal.

(See *Lesson 6*)

In the earthworm castings, the soil, having passed through the digestive apparatus of the earthworm, has been thoroughly triturated. This treatment of the soil breaks down most, if not all, of the chemical and mineral elements and properties. It makes them more easily available to the roots of the grain, adding to the quality of the roots and sprouts.

The container in which Intensive Range is prepared should be filled with a mixture of—

- One third castings
- One third peat moss (or substitute)
- One third grain

The "Trays" are then set aside and kept moist until the grain sprouts, which takes from seven to ten days, depending on the temperature.

In cold weather, the "trays" may be artificially warmed, making sure the grain germs heat enough to germinate and grow. Care should be taken not to overheat the Intensive Range.

When the grain has sprouted, the contents should be placed outside the hennery, but within reach of the fowls.

The writer considers one basin of Intensive Range sufficient to feed twenty-five hens for two days. Once the poultry raiser has prepared a few basins of Intensive Range, he will find that a basin can be prepared in less than three minutes.

In addition to Intensive Range, poultry should be fed (at night) dry grain (preferably wheat), about one pint to twenty hens. One must regulate the amount of grain according to the fatness of the fowls. Do not allow them to get too fat. Fat hens are poor layers.

Grit should also be supplied, preferably shells which supply a goodly portion of lime.

Persons with small flocks will, in a short time, keep sufficient "trays" under preparation, so that when one "tray" is consumed another will be ready to take its place, thus keeping the chickens constantly supplied with live food.

All ages of poultry will benefit from Intensive Range, from very young—for whom this method is exceptionally beneficial—to very old.

For larger poultry farms, Intensive Range is prepared inside the enclosure of the hennery by utilizing the scratching area.

While the birds (divided according to your personal idea, from ten to twenty-five chickens to a unit) are cooped in one side or compartment (see preceding lesson), the grain is sprouted in the other.

Here we follow the same general principle as when preparing Intensive Range in "trays." The scratching area, being impregnated with earthworms and their castings from the pits under the roosts, is permanently ready to be planted in grain. The amount planted here is governed entirely by the area the poultryman wishes to cover.

Grain planted in hennery soil will, in warm weather, take root and sprout within ten days. In colder weather, it may be necessary to cover the hennery with canvas. This is something that should be done in all severely inclement weather.

We shall now move into *Lesson 6*, where we shall observe the part the bluebottle fly plays in Intensive Range when properly put to work by poultrymen who seek sound profits. There, too, we shall observe further facts regarding food for poultry, intelligent management of the poultry farm, and a few important points that should be impressed on the minds of anyone interested in receiving profitable returns from an investment in henneries, stock and feed.

> *Learn to keep the door shut, keep out of your mind*
> *and out of your world, every element that seeks*
> *admittance with no definite helpful end in view.*
>
> —*George Matthew Adams*

Putting the Bluebottle Fly to Work

▚

*The truth about flies being enemies of man —
J. Henri Fabre quoted — Subject of fly larvae
should not be considered obnoxious — Potency
of the house fly — Pepsin — Fly larvae high
in this necessary element — Preparing trays
to trap the larvae — How to encourage the
"blow" where it will be beneficial for poultry
— This system is sanitary, and, when properly
operated, will very likely rid the poultryman's
house of annoying "blowflies"*

Countless ages before *Homo sapiens* evolved from some yet undetermined anthropoid, various small, two-winged insects flitted about heaps of decaying animal and vegetable matter. Among these were insects which we have come to call flies.

To the ordinary person of so-called civilized areas of the earth, the fly is a pest to be ensnared on sticky paper, trapped, swatted or cursed at according to individual inclination.

Thoughtless persons might imagine that if flies were banished from the earth, this swirling sphere would take a long stride toward becoming a paradisical place on which to live. Actually, if such an extermination were humanly possible, which it is not, instead of the earth becoming a paradise, it would quite likely become much more of a hell than it is now claimed to be by pessimists.

Since modern medicine became so interested in public health and sanitation, we have been taught, from earliest school days, that the fly is an abomination, a bearer of disease, epidemics and death.

With no desire to "stick out my neck," I shall not deny the element of truth found in those statements. But I do object to any system that offers through our educational channels a half-truth. And to claim that the fly is only a bearer of sickness and disease is a half-truth.

Science is quite cognizant of the good work the fly performs in life; a good work it was performing long before man acquired the ability to use or misuse drugs and surgical instruments.

> "The larva of the fly is a power in this world. To give back to life, with all speed, the remains of that which has lived, it macerates and condenses corpses, distilling them into an essence wherewith the earth, the plant's foster-mother, may be nourished and enriched...(There are hosts of these larva) to purge the earth of death's impurities and cause deceased animal matter to be once more numbered among the treasures of life," Jean Fabre declares in his remarkable book, *The Life of the Fly.*

Jean Henri Fabre (Faw-br) was born in France, December 21, 1823, and died October 11, 1915. He was a schoolmaster, Professor of Philosophy and a scientist of immortal note. His literary works include, *The Life of the Fly*—which every reader of this volume should read—*Insect Life; The Life of the Spider; The Mason Bees; Bramble Bees and Others; The Hunting Wasps; The Life of the Caterpillar; The Life of the Grasshopper; The Sacred Beetle and Others; The Mason Wasps; The Glow-Worm and Other Beetles; More Hunting Wasps; The Life of the Weevil; More Beetles* and *The Life of the Scorpion.*

Fabre's works show a minute and sympathetic observation of the habits and lives of insects. Charles Darwin was a sincere admirer of Fabre and bestowed upon him the title of the "incomparable observer." Fabre wrote in a friendly, intimate and absorbing style.

As the reader may have observed, a few references have been made to the fly in previous chapters. I purposely did not go into detail regarding this insect and the part poultrymen may easily make it play (and pay) in producing better poultry. I saved what I have to say about the fly and its eggs and larvae for this chapter. Here I shall devote sufficient time and space to the subject without confusing it with other matters.

For the past two decades I have found it practically impossible to discuss this matter without having the subject changed by listeners. As if speaking about flies and their eggs was a forbidden subject.

This disinterested reception of what I know to be beneficial facts (substantiated by reputable men of science) is discouraging. Many times I have resolved to stop trying to point out the value of the fly to poultrymen. It seemed to me to be a waste of valuable time trying to interest apathetic individuals in putting the fly, particularly the bluebottle, to work.

I have realized for many years that this subject is avoided because of the wholly black reputation given the fly. It is true that flies can, and do, carry bacteria on their feet and the hairs of their legs and bodies. That these bacteria have, and do, infect humans is also true. But accusing the fly as the prime suspect in this matter is merely excusing the ignorance, carelessness and apathy of the general public.

The healthy human body, nourished by non-deficient foods, is quite capable of throwing off the bacteria brought to it by the fly or any other medium. It is the weakened body, the poorly and improperly-fed body, which is susceptible to illness. Such bodies are fertile breeding grounds for disease bacteria.

However, in the fact of this general attitude toward the fly, her eggs and larvae, I am determined to approach the subject without further detours. No matter how much the reader may object to this subject, he cannot interrupt what I have to say. Of course, he may set aside this volume or skip this chapter. He can very easily close his eyes and mind to the facts I have gathered here. After all, it is up to the reader whether or not he wants to increase his knowledge and raise better poultry by putting the bluebottle fly to work.

Flies, like all other living things, are divided by naturalists into species. These cover a wide variety, extending from the small, gnat-like fly to insects very closely related to beetles.

Our concern is especially with the common housefly, of which there are a number of species. Common names for these are: blowflies, horseflies, fleshflies, screwflies, gray fleshflies, greenbottles and bluebottles.

All of these are classified by science as *Luciliae*. Loosely defined, this means "flies that glitter." These are all familiar to us, even if we cannot distinguish one from another.

Reclassed for our purpose, I divide these flies into three groups:

a) Greenbottle
b) Gray fleshflies
c) Bluebottles

The first two species are not given to haunting our kitchens, though both of them do pay occasional visits to the culinary department when attracted by strong cooking odors.

It is the bluebottle that pillages our poorly protected meals. She is the "blowfly" whom all housewives abhor.

However, both the greenbottles and the gray fleshflies also "blow." That is to say, they select a suitable piece of decaying animal or vegetable matter and, in some dark recess, "blow" the contents of their ovaries into it.

During this egg-laying period, which varies between species, each fly lays from 100 to 600 eggs, or, as in the case of the fleshfly, hundreds of live larvae or grubs.

One mathematically-minded writer has shown that the progeny of one housefly—from May 1st to September 30th—will number three trillion, nine hundred and eighty-five billion, nine hundred and sixty one million, seven hundred and fifty-five thousand, one hundred (3,985,961,755,100).

In the case of a bluebottle—henceforth we shall dismiss the gray fleshfly and the greenbottle—the eggs hatch in about two days, if the weather is warm.

From these eggs appear small worms, 0.029 of an inch in length. In a wriggling mass, the worms instantly begin to seek nourishment.

> "They do not eat, in the strict sense of the word," says Fabre, "they do not tear their food nor chew it by means of implements of mastication. Their mouthparts do not lend themselves to this sort of work."

Through their oral openings, the larvae excrete saliva that instantly liquifies the matter it contacts, turning it into a broth. This they drink.

Digestion is, after all, a form of liquification. In the instance of the bluebottle larva, it digests its food before swallowing it.

This curious form of exterior digestion is of vital interest to all poultrymen.

Pepsin, which fowls have no way of preparing within themselves, is the chief necessity of adequate digestion. Fowls, especially those raised as meat birds, should be fed foods containing pepsin.

Commercial pepsin is scraped from the membranes of the stomach of the pig and the sheep. Yet, as Fabre points out, commercial chemists "would obtain a product of the highest quality" from the larvae of the bluebottle, for these grubs produce a "pepsin of singularly active kind."

From the moment the larva breaks from its egg, it begins to eat its way through approximately ten days before it enters the dormant cocoon period.

As the larva approaches the cocoon period, it departs from the home it has had since birth. Under favorable circumstances it leaves its surroundings and buries itself in loose soil, down just a few inches. There it remains until it breaks free of its cocoon, and, after adjusting itself physically to its new life, flutters its wings and goes about its life as a full-grown bluebottle.

It is the embryonic stage—that of the larva between the egg and the cocoon—in which we are especially interested.

Not only is this grub a delicacy for fowls, both domesticated and game, but it is extremely high in pepsin, as I have already stated.

However, the poultryman who wishes to put the bluebottle fly to work for him should not imagine that he is the first to do this.

Many years ago, caretakers of large European estates fattened fowls, domesticated and game, by a method of feeding fly larvae to the birds.

The system employed by the caretakers, which may be employed today by poultrymen, was to hang a piece of meat above a tray of meal. The meat was hung high enough to permit the air draft to carry off the odor. The flies "blew" this meat, the eggs hatched, the larvae grew and, in due time, fell into the tray of meal. Here, the birds snatched them eagerly.

In putting the bluebottle fly to work, poultrymen with small flocks may attract the flies to their Intensive Range "trays" by placing fish gills, other fish offal, or decaying offal of any kind upon the top of the "trays."

These trays may be so placed as to avoid annoying either the neighbors or members of one's family by the odor. Be assured that the flies will find such "trays" and will "blow" them.

Properly-managed attractions for flies may actually rid the poultryman's home of "blowflies." If objectionable decaying matter is easily obtainable in the "trays," the flies will desert the kitchen.

My preference in preparing magnets to attract the bluebottle so that the birds will receive the utmost from this type of food supply is as follows:

1) For small flocks, prepare the "trays" as above.
2) For large flocks, follow the system used by European caretakers of poultry.

The latter is best done by suspending the decaying matter near the ceiling of the hennery. This should be done so that there will be an air draft that will carry off the odor. Odors from decaying matter are gaseous, and, being lighter than air, will rise and dissipate in the open.

Little if any labor is required to keep this larvae-breeding matter perpetually functioning. When the decaying matter is hung, the flies will locate it and Nature will take her course. When the larvae have matured to a degree where they are ready to enter the dormant stage, that period during which the larvae is transformed from worms to full-grown flies, they will drop into the tray set directly beneath the decaying matter.

The genuinely sincere believer in this system may carry on the feeding of larvae throughout the year by constructing a fly-house. By keeping the house adequately heated in cold weather, properly sealed, though admitting sunlight, bluebottles may be successfully raised.

From a sanitary viewpoint, this system should meet with general approval for it segregates and controls the fly larvae. One of the objections invariably brought to my attention is that this system of feeding live food to poultry will breed flies.

No statement can be further from the truth. Instead of breeding flies, it actually reduces their number.

It does encourage the breeding of larvae, but fly larvae are not flies.

In conclusion, I wish to emphasize one paramount point— money can be made in poultry *if* the poultryman will, first, *feed live food to his birds*, and second, *reduce his cost of feeding them*.

I do not claim that my system of Intensive Range is the acme of poultry feeding systems, but I do claim that a poultryman can develop it to a point where he may rightly claim it perfect.

Part 3

The power to think, consecutively and deeply and clearly, is an avowed and deadly enemy to mistakes and blunders, superstitions, unscientific theories, irrational beliefs, unbridled enthusiasm and fanaticism.

—*Haddock*

Introduction

∷

Individuals who engage in the business of breeding, raising and selling earthworms are, in a general sense, pioneering. But in no sense are they experimenting. The experimental labor has been done for them. All the prospective earthworm farmer need do is follow the advice and suggestions of those who have paved the way for him in this new and interesting business.

Once again I repeat—do not enter the earthworm farming industry in the hope of becoming a plutocrat over night; and do not become interested in this business in the manner of a five-year-old girl playing house.

The business of breeding, raising and selling earthworms is a reputable, sound and profitable business. It is one that must be operated along regular business principles with business acumen if success is to be attained.

A successful businessman, be he seller of locomotives or children's tricycles, is the individual who knows his business. The high-geared competitive system of the present era demands more from businessmen than any other era in the history of industry and commerce. And the man who cannot keep pace with modern competition will soon be bringing up the rear, like the tail of Mrs. Casey's cow.

As we shall see in this volume, a prospective, ambitious earthworm farmer does not need to make more than a comparatively-small investment in preparing his working equipment. But he, or she, must study the business of earthworm farming, both from its production and its selling angles.

Probably the most effective manner for the writer to emphasize these two points—inexpensive equipment and a studious interest in earthworms—is to briefly review the manner in which the immortal Charles Darwin worked during his experiments.

Darwin may be rightly called the discoverer of the indispensability of the earthworm as an aid to soil productivity. He believed far more in acquiring knowledge than he did in the quality of his equipment.

Darwin's instruments were of the simplest known. In his day (1809-1882) the compound microscope was still a weak and undeveloped instrument. He used a microscope, probably not as good, nor as accurate, as any of those that may be purchased today for a few dollars.

The great naturalist considered tool and instrument makers of his time and place infallible, yet his micrometers varied greatly from one another. For lineal measurements he used an ordinary three-foot rule, which was

> "...the common property of the household, and was
> constantly borrowed, because it was the only one
> which was certain to be in its place—unless, indeed,
> the last borrower had forgotten to put it back."

For measuring tall plants, shrubs and young trees, Darwin used a seven-foot pole which was graduated by the village carpenter. No mention is made of what the carpenter used as a standard! The results of Darwin's work, unequaled in his particular scientific field, obviously proves that the adage, "A good workman is judged by his tools," like all other rules, has its exceptions.

But, no matter how mediocre Darwin's equipment was, his method of keeping records of his experiments was in complete contrast to his carelessness concerning his tools and instruments.

His measuring rule may have been a fraction of an inch inaccurate, his microscopic lens weak, his micrometers loose and uncertain, but the notes Darwin made, the time he spent mentally preparing his material which was to be given to posterity through the media of papers, pamphlets and books, required labor with tools that had to be accurate—and those tools were words.

But let us turn to Darwin himself:

> "I will add that with my large books I spend a good deal of time over the general arrangement of the matter. I first make the rudest outline in two or three pages, and then a larger one of several pages, a few words or one word standing for a whole discussion or series of facts. Each one of these headings is again enlarged and often transferred before I begin to write in extenso. As in several of my large books facts observed by others have been very extensively used, and as I have always had several quite distinct subjects in hand at the same time, I may mention that I keep from thirty to forty large portfolios, in cabinets with labeled shelves, into which I can at once put a detached reference or memorandum. I have bought many books, and at their ends I make an index of all the facts that concern my work: or, if the book is not my own, write out an abstract, and of such abstracts I have a large drawer full. Before beginning on any subject I look to all the short indexes and make a general and classified index, and by taking the one or more proper portfolios I have all the information collected during my life ready for use."

Wise, indeed, is the individual who will systematize his knowledge as did Darwin, and from the above quotation the prospective earthworm farmer should receive lasting help.

These two characteristics of Darwin were purposely brought in here to impress upon the reader that equipment need be neither expensive nor ostentatious.

What counts is studious attention to the earthworms and their habits. This makes for success.

All the need-to-know points which the earthworm farmer should understand are made clear within these pages. Neither I, personally, nor any other individual can make a success of an earthworm farmer. The quality to be successful lies within the individual engaged in earthworm production.

And so, with these friendly pointers. I bring you now the fundamentals of earthworm farming, leaving this introduction with just one more repeated, important sentence—the average man or woman can become a financial success through breeding, raising and selling earthworms if he or she will wholeheartedly apply known and proved facts to the business.

Lesson 1

Natural and Man-made Enemies of the Earthworm

■■

Every living thing has natural enemies — U. S. Department of Agriculture quoted regarding enemies of earthworms — The Argentine and Pharoah ants — Greatest enemies — How to defeat them — Earthworm farmer has little to worry about, except for these ants — Be careful of Cyanide in any form — Anything that kills earthworms should be shunned by all farmers, nurserymen and gardeners

Like all living things, the earthworm has a long list of enemies. It is distressingly unequipped to "fight back," is minus all trace of either defensive or offensive apparatus, and would, if left alone, be a model of pacifism and isolation. Its complete cycle of existence is devoted completely to seeking nourishment for itself and propagating its kind. It has no interest in anything else.

We are not especially interested in the natural enemies of the earthworm as they apply to the earthworms under the watchful eye of the earthworm farmer. It is a good idea to consider this topic briefly.

The writer feels this is best done by turning to the United States Department of Agriculture's Bulletin, Number 1569, from which the following is taken:

"In addition to the fisherman with his ever-ready garden fork and tin can, the earthworm has numerous natural enemies that are constantly alert. Song and game birds, as well as domestic poultry, patrol the meadows and woodlands during the day, while the ubiquitous toad with his huge appetite and sticky, darting tongue, stalks earthworms during the early hours of spring and summer nights. Some of the smaller species of harmless snakes feed largely on earthworms at times. The Bureau of Biological Survey has identified earthworms in the stomach of no fewer than 45 species of birds. The crow seems to get more egg capsules of the worms, and the jacksnipe, woodcock, starling and robin seem to get more adult worms. Earthworms are an important item of food to the four species named.

This warfare, however, is not confined to the mere surface of the soil. The earthworms are by no means safe even in the fastnesses of their underground burrows. Here they are pursued relentlessly by the voracious shrews and moles, which subsist largely upon these creatures. Centipedes and 'thousand leggers' follow the worms even to the depths of their longest tunnels.

In addition to these formidable foes, some kinds of earthworms have still another enemy in the form of a two-legged fly, which superficially resembles closely the common house fly (*Musca domestica*). In point of fact this insect, which is known as the cluster fly—*Pollenia rudis*—and often enters houses in large numbers in the fall, is commonly mistaken for the house fly. By the use of a low-powered magnifying glass, however, the cluster fly is easily distinguishable from the house fly as it has a downy coat of yellow hairs upon its back and sides. This down is entirely absent in the house fly.

For a long time the manner of life of the cluster fly was unknown, but this problem was solved in 1908, by David Keilin, of the Quick Laboratory, at

Cambridge, England. He found that the larvae or maggot of this fly was parasitic upon certain species of earthworms. The fly's eggs are deposited on or in the soil and hatch in from five to seven days. The minute maggots then seek the worms and bore into their bodies, where they continue to feed until eventually the worms are destroyed. Subsequent investigations carried on by American entomologists confirmed in a general way these facts as applied to the life history of the insect in this country, but it was found that in summer the eggs hatch in about three days. It was found, also, that there may be several generations of the fly in a year and that it is possible for the insect, under certain conditions, to undergo its complete cycle of development in from twenty-seven to thirty-nine days. The cluster fly itself falls a victim in great numbers to a fungus disease which attacks it during the warm, humid weather of mid-summer. At this time even the newly-emerged flies may be observed, stilled in life-like poses, attached to the leaves of plants, where they have been killed by the action of this fungous enemy."

Not mentioned in the previous excerpt, though probably the earthworm's greatest enemy in farms, nurseries, truck patches and gardens is the ant or emmet.

In the development of the domesticated earthworm for commercial use, it was definitely proved that, various species of ants caused problems. Experimentalists worked very hard to defeat the destructive ability of both the grease-eating and the sugar-eating ants. Numerous insecticides were used, some with no results, others with fair results. Only one commercial insecticide out of nearly one hundred was found that proved itself a definite destroyer of the two most pernicious species of ants on the West Coast—the Argentine, or sugar-eating ant, and the Pharoah, or grease-eating ant.

As we have seen in *Lesson 2* of *Part 1*, the earthworm's digestive system is highly sensitive and a prey to environment. While its system absorbs the needed food energies from the large quantities of soil it swallows, the earthworm shows a preponderance of favor for fats and a general interest in sugars. These foods may rightly be called the staples of the earthworm's diet.

Since all ants are lovers of fatty or greasy foods, it is readily understandable how an ant colony may eat or otherwise dispose of the fats in the soil in which earthworms are dwelling. A colony of ants, being energetic and tireless workers, will, in a comparatively short time, devoid the soil of fats, thus taking from it one of the chief food necessities of the earthworm.

The Argentine ant, existing in large numbers in and about the southern portion of the West Coast, is the only sugar-eating ant of the seven species catalogued in that section. In northern Mexico and many sections of the United States southwest, the Argentine ant is known as the "honey ant." It lives exclusively on sweets and is one of the hardest insects to destroy through the media of insecticides.

In developing, breeding and culturing earthworms for commercial use, the ant should be considered their most dangerous and destructive enemy. With culture beds always under control by the earthworm farmer, he has little to worry about from the general run of earthworm enemies—except the ant.

Though largely carnivorous, frequently fatally attacking animals much larger than themselves, ants do not prey directly on the bodies of the earthworms. Principally because the annelid spends most of its time beneath the surface of the soil. But because the ant does not attack the earthworm directly does not reduce its ability to destroy it indirectly.

Experiments were carried on for many months by California earthworm farmers regarding the potency of ants as denuders of the soil of sugars and fats. These experiments were performed in carefully-secluded and prepared earthworm culture beds. The work of the Argentine and Pharoah ants was painstakingly watched, checked and recorded. In all instances, the energetic ants denuded the soil of sugars and fats to such an extent that the earthworms began to show a decided decline in activity. As the worms began to weaken and die, the experimentalists gathered the remaining worms and placed them in culture beds free from ants.

The results of this change were immediately noticeable. Able to secure the needed starches and desired sugars, the earthworms showed instant signs of gaining a new lease on life. In a few days, all trace of the lethargy, generated by the deficiency in sugars and fats in their food, disappeared.

It was while these experiments were in progress that efforts were made to procure an insecticide that would satisfactorily discourage or completely destroy these enemies of the earthworm.

It was found that insecticides definitely destructive to certain species of ants did not kill other species. Since all ants are enemies of the earthworms, it was essential that a suitable and potent insecticide should either be found or developed.

In time, a West Coast insecticide was found that did satisfy the experimenters. Of the seven known species of ants on the West Coast, none were able to withstand the new concoction.

The user of insecticides, and other poisonous solutions or powders to destroy such pests as shrew, moles and centipedes (all enemies of the earthworm), should employ judgment in placing the poison.

Most poisons used by farmers and orchardists to rid their premises of pests contain cyanide in one form or another. The earthworm breeder who promiscuously spreads any of these should realize that such lethal preparations, sufficiently potent to kill insects and animals, will also kill his worms, were they to come into contact with them.

Moles and other small burrowing animals that are enemies of the earthworm, may be safely destroyed by sprinkling poisoned grain at the entrance of the burrows. In this way, it is usually beyond the reach of the earthworms. In areas where moles abide, earthworms are known to penetrate deeper into the soil. By keeping as far away from the moles as it can, the earthworm is clearly expressing a natural law—self-preservation.

Under no conditions should the breeder of earthworms attempt to destroy enemies of earthworms without realizing that what will kill the goose will also kill the gander—a poison that will kill a mole will likewise kill an earthworm.

And so, I close this lesson with an emphatic warning—be careful when you begin to rid your earthworm culture bed of enemies.

If you are not careful, you may destroy both the enemy and the friend.

The Trout Farmer's Problem

::

*Raising fish in captivity age-old art —
Approved today by national, state and private
institutions — Original feed for fish in captivity
— Change to by-products of meat and fish
packing concerns — How pet food threatens
this industry — Remarks regarding pet food
by well-known bio-chemical analyst — Advice
to cat and dog owners — Possibilities of the
earthworm as food for fish in captivity*

Following the course of human industrial and commercial progress, there is always a wake of disturbing conditions, causing personal unrest, uncertainty and financial loss. In time, these disturbances mold or blend, readjusting to meet the change. Sometimes the new mode is introduced quickly; in other instances, many years pass before equilibrium is reached.

Here is an example to clarify this point.

With the advent of the automobile, the horse, mule, harness accessory, carriage and wagon industries suffered greatly. Many of these businesses suspended operation; others blended amicably with the new conditions brought about by the "horseless carriage."

Today, one of the oldest arts in the history of man faces the grim outlook of being caught in one of these wakes of the good ship progress.

The writer refers to the art of pisciculture—feeding and fattening, breeding, rearing and preservation of fish by artificial means.

Much older, indeed, is this art than the ordinary person imagines. It dates back to the central period of ancient Egypt. Later, across the Mediterranean Sea, the Greeks practiced it, and halfway around the globe from Greece, the Chinese reached a high degree in its development.

During the past century, pisciculture has shown remarkable strides, especially in America. Its greatest forward step was taken when science discovered that the ova of the fish could be taken from the female, impregnated with male milt, and hatched in tanks. This took away much of the element of chance (customary under natural spawning conditions) that the milt would fertilize the maximum number of eggs.

For the past quarter of a century, the art of artificially-raising fresh water fish—though the artificial culture of oysters, clams, mussels, lobsters and other crustacea is also profitably practiced—has come into general favor.

Our national government, under the American Fish Commission, has deeply interested itself in this art. It is practiced in the United States, Canada and many European countries on a surprisingly large scale. Many of our states, too, have piscicultural departments and there are thousands of "fish farmers" scattered throughout the nation.

With national and state governments taking a thorough interest in this art, it has shown great improvement and has become an important department of our commercial life. Millions and millions of fish are "planted" annually by our governments and privately-owned piscicultural institutions.

And now, believe it or not, this productive industry is facing a possibly insuperable barrier.

Curiously enough, cats and dogs are innocently responsible for the clouds of doom that are gathering above the art of pisciculture.

Cats, dogs, and their owners, who seem to fall easy victims to anything and everything that saves them a few minutes work.

During the past decade, the fastest growing industry in the nation has been that of packing and canning pet food. So enormous has become the sale of these prepared pet foods that their manufacturers are buying practically all of the by-products of meat packing houses and fish canning concerns. In addition to this, they are more or less responsible for the slaughter of wild horses on the ranges of our northwestern plains.

The Rainbow Angling Club and Hatchery discussed this problem in 1938. They declared that national, state and privately owned fish hatcheries will have to come up with something really creative if their managers and owners cannot do something about feeding their fish. Especially if the trend toward greater sale of pet foods continues.

On the surface, all of this may sound like material for Robert Ripley. Yet when the matter is analyzed, it is not as fantastic as it at first appears.

Let us consider the food proposition as faced by the Rainbow Angling Club and Hatchery and see how cats and dogs, indirectly, were causing them no little amount of worry.

The Rainbow Hatchery deals exclusively in rainbow trout. The hatchery and propagation plant is located about fifteen miles from the source of Mill Creek at springs high in the San Bernardino Mountains in Southern California. Mill Creek supplies 18,000 gallons of mountain water every minute to the hatchery, in which there are one and one-half million trout.

To this piscatorial metropolis, 7,500 pounds of ground meat and fish meal, a similar amount of cereals, are necessary every week. This is about 780,000 pounds annually, or approximately 400 tons, half of which (about 200 tons) is ground meat.

There are few, if any, streams in the world capable of caring for 1,500,000 trout; certainly none that could supply this number with adequate food the year around.

Fish in their natural state feed on other small fish, insects or insect larvae. In captivity, under artificial conditions, such natural food is not available in an amount even vaguely bordering on sufficiency. The only substitute known with enough food energy are the by-products from meat and fish packers.

For many years, experiments have been carried on in the hope of finding a suitable substitute for fish food other than these by-products. To date, no success can be reported. Ground meat—made from these same meat by-products used in pet foods—contains vitamins and proteins necessary for a balanced diet for fish.

The feeding of cereals to fish in captivity is done merely to "fill" the fish. Cereals have no food value as far as fish are concerned. Being cold blooded vertebrates, they have no need for sugars or starches.

Fish food must be of the proper quality and the proper quantity if the hatchery man would successfully raise fish. The water, always running, should be between 40 and 60 degrees Fahrenheit.

The "range," that is, the number of fish to a given area, must be such as to avoid crowding. Crowding is dangerous, retarding growth and spreading disease.

Young fish are fed frequently, often as many as ten times a day. As they grow, this frequency is tapered off until, at maturity, they are fed but once a day.

> "When I first became engaged in the pisciculture art, the staple food for fish was liver," they told me. "But the public was told that liver was beneficial for them. They were advised to eat it—and the price of liver rose to such a figure that we could no longer afford to feed it to fish.
>
> It was then we turned to the general run of by-products of the meat packing houses. Here we found we had an ever-present and seemingly unending supply of food ideally suited for fish in captivity. Then came the pet foods. Companies making them began to buy more and more of the meat packers' by-products. In a few years, the demand began to run

nip and tuck with the supply and the meat packers boosted the price of their by-products—by-products that a few short years before were a drag on the market.

Our difficulties are now becoming acute. Not only is it becoming more and more difficult to buy these by-products, but the price has risen to a point that is dangerously close to a figure too high for our consideration. At the present time, I can see no signs that the price of this commodity will come down—which is fine for the meat packers (and I hope, for the farmer!) but it is certainly not a pleasant picture for fish hatchery owners to visualize.

Here on the West Coast, there is one pet food manufacturer who has five boats plying Pacific harbors buying all the by-products from the fish canneries available."

It might not be amiss to insert here a reference foreign to the subject of fish hatcheries, yet of sufficient import to readers of this part who have a pet cat or dog to warrant its inclusion.

Many of these pet foods bear the phrase "government tested" on their labels. Discussing this phrase, J. W. Patton, experimental biologist of East Lansing, Michigan, says:

"...This phrase means one thing on a dog food canned in an establishment in which Federal inspection is maintained, and canned under the same conditions that exist for canning meat; but quite another on scrap and offal from inspected animals...'Made from U. S. Inspected Meat,' 'U. S. Inspected and Passed'—passed for what? Edible or inedible? Food or fertilizer?"

Such phrases may be on a can that "may contain 50 percent lung and 50 percent weasand handled like coal...". "Made from U. S. inspected meat" is a phrase openly used by unscrupulous manufacturers of the most unworthy feeds on the market; not only to cover up their inferiority, but to imply superiority—and the dog and cat suffer.

Such food manufacturers, Mr. Patton claims,

> "may purchase scrap or even fertilizer from inspected
> establishments and use it without restriction."

Such firms, we are informed, mix the foregoing with

> "meat from a desiccating works and still advertise
> that they are using U. S. inspected meat. Bones and
> trimmings from the meat market, although they may
> go to a rendering plant and are shipped to a pet food
> manufacturer, are still 'U. S. Inspected and Passed.'
> True enough, they were formerly inspected and
> passed, and at the time were clean and wholesome...
> There is no authority in the meat inspection law to
> control them (the pet food manufacturers)."

And this is the business that threatens the very existence of fish hatchery operators!

"What are we going to feed our fish when the pet food industry monopolizes the meat and fish by-products which are necessary food for fish in captivity?" is a question thousands of fish hatchery operators in the United States are asking today.

Elsewhere in this work, the reader has been presented with the advantages derived from feeding earthworms, through Intensive Range, to poultry and game birds.

It is the writer's personal conviction that properly-fed and properly-controlled earthworms, in properly-prepared beds, may produce the necessary food elements required by growing fish.

Earthworms may become a new and lasting substitute food for fish in captivity.

The attention of the reader interested in this feeding phase is called to the lesson *Putting the Bluebottle Fly to Work* (p.78).

Similar methods to those used to supply hens with live larvae may also be employed by fish farmers to their advantage, and also by frog farmers, as we shall see in the next lesson.

The only objection raised to this form of live food feeding seems to be the resultant odor of the matter used to attract the bluebottle.

Even so, both the fish and frog farmers should be able to select a place for the bluebottle to "blow" that will not be an offense to the olfactory nerves of humans in the vicinity.

A complete description of preparing the tray to hold the matter attractive to the bluebottle, and the best place to suspend the tray, will be found in the following lesson.

If the reader is interested in pisciculture, he must realize the enormous quantities of earthworms that will be required to care for the appetites of large hatcheries.

This, though, is not a handicap, for the writer knows that earthworms can be produced in sufficient quantities to meet the demand of hatchery owners.

Readers who would like more information on this subject should know about the progress of the current experiments carried on by the writer. I hope, in the space of a few months, to be able to produce a small booklet covering these experiments and this I shall make available at a price that will cover the actual cost of the booklet.

I would suggest in closing that persons interested in this subject drop the writer a card. The names of such persons will be carefully filed and the writers advised when the experiments are concluded, the results tabulated and the booklet ready for distribution.

And now let us glance at some of the problems of the frog farmer.

Feeding Problems of the Frog Farmer

::

"Money in frogs" — List of frogs that may be raised commercially — Natural food for frogs — A. G. Peek, Southern California frog man, quoted — What frog food should be used — What its qualifications should be — Frog farmer should raise earthworms from egg capsules — Feeding basket — Master bed — Discussion of electroculture for use by frog farmers

Let me lay my cards face up on the table before I start dealing advice and suggestions in this lesson. I do not wish to encourage or discourage an investment of time or money in the business of frog farming for commercial gain.

I am quite cognizant of the fact that frog farming is, on the one hand, a legitimate business, scientifically taught. On the other hand, it is a racket sponsored by empirical promoters.

Be that as it may, I shall not become dictatorial in the matter. It should be up to the individual thinking of entering the frog farming business to judge whether or not he is dealing with a reputable firm.

"Money in Frogs" (and similarly-attractive advertising catch lines) is ambiguous, regardless of the fact that such statements are acceptable as good English. The fact is, there is no money in frogs, nor earthworms, nor cats, nor horses, nor vegetables—without that form of human energy which we call "work."

It may seem trite to declare that financial gains are produced by the individual raising these commodities. The commodities are the vehicles on which human energy rides to financial profit. This point is well worth emphasizing. Many persons are led by advertisements to believe that all they have to do to make money is to stock a farm with this or that and money will begin to pile up.

I have previously stated in this work—in discussing the raising of earthworms for commercial use—that it is foolhardy to enter any business with the hallucination of wealth over night. Such hopes are the fabric from which fairy tales are spun—and fairy tales are further removed from fact than is fiction.

Nonetheless, a goodly annual income can be developed through the business of frog farming. For those already engaged in this business, and for those contemplating entering it, this lesson is especially prepared.

The frog farmer's problems of feeding differ from those of the fish farmer. The latter, as we have already seen, is facing a food shortage for his piscatorial school. The frog farmer's problems also concern food, though they are not as acute as those of the fish farmer.

If, as some dealers in frogs claim, the frog farmer believes he can go to the nearest abattoir and buy meat scraps for a mere nothing, he had better prepare himself for a shock. The previous lesson explains that matter and shows where the by-products of meat packing houses go.

Breeding, raising and fattening frogs—of which there are four kinds suitable for commercial raising in the United States, i.e. French, Chinese, Louisiana and American—in captivity is definitely artificial. Such frogs, if they are to increase and grow to marketable size, must be fed by the farmer. Like fish in captivity, frogs cannot produce sufficient food in enclosed ponds without human assistance.

Frogs, in their natural state, live on insects, insect larvae, various types of worms and crawfish (crayfish). During the tadpole stage, they are, for the most part, vegetarians.

But I have not set out to discuss the dietary habits or needs of frogs in captivity, except as related to earthworms and the larvae of the bluebottle fly. Your local library should have a number of books on this subject, or your Chamber of Commerce. Even the Board of Trade should be well enough informed about this matter to suggest a frog-canning concern to which you could address your inquiry regarding frog feeding. Most, if not all, of these agencies will be pleased to send you suggestions.

My primary interest in this series is our friend, the earthworm, and, in this lesson, the part it plays as a food supplement for frogs in captivity.

A. C. Peek, an official of the Rio Hondo Trout and Frog Farm, El Monte, California, in discussing frog food with the writer, declared that

> "Earthworms are a beneficial delicacy for frogs. They
> eat them ravenously, both in and out of captivity, and
> they should be an integral part of the frogs' diet."

Frog food should have these qualifications:

1) It must be abundant.

2) It must not be injurious to land or marine vegetation, insects, or other natural food for frogs.

3) It must be low in original cost and upkeep.

4) It must be of such dimensions that all frogs can eat it.

5) It must be capable of withstanding changes in temperature, ranging from summer heat to winter cold.

6) It must be available for frogs where and when they want it.

Earthworms meet each of these six requirements thusly:

1) they are prolific breeders

2) they will not injure plant, animal or insect life

3) their cost and upkeep is low

4) they are never too large for even the youngest, smallest frog

5) they will withstand various temperatures, when properly housed; and

6) they are made available as frog food with very little labor on the part of the farmer.

Because earthworms are slaves to the environment in which they have been raised, I do not look with glowing favor upon suggestions made by some teachers of frog culture who advise frog farmers to dig their own earthworms and transplant them to a previously-prepared culture bed. This system is not always successful. At best, it is a slow process. The earthworms that withstand the change of environment will be slow breeders, much too slow for the needs of the frog farmer.

My theory, substantiated by many tests over a period of years, is that the culture of earthworms, regardless of what their future use may be, should begin with the egg capsules. In this way, the newly-hatched earthworms will immediately partake of the nourishment available. They will mature naturally, in about ninety days, and when properly housed, controlled and fed, will reproduce regularly every week throughout the year.

I have two suggestions to make regarding the earthworm culture bed for frog farmers.

First: For large frog farms, a culture bed not to exceed twenty-five feet long, ten feet wide and four feet deep should be excavated. The location should be shady, and, if possible, damp.

Fill this bed (about six to eight inches below the top) with a mixture of peat moss (or sawdust or wood shavings), rich, black soil, manure from grain-fed horses or cattle, chicken droppings, and about a bushel of leaf mould. No care should be taken in filling the bed, except, of course, to have the ingredients well mixed.

Into this bed, I would place 50,000 earthworm egg capsules and allow nature to take her course.

In two months, if the minimum amount of earthworms and egg capsules are placed in the bed, the frog farmer may begin to harvest his crop of annelids for frog food.

Care should be taken during the daily, biweekly, or weekly harvesting of the earthworms that too many breeding earthworms are not fed to the frogs. Such carelessness would soon deplete the earthworm culture bed of its most valued asset—prolific breeders. (Elsewhere in this work, the culling of young, immature and sluggish, earthworms from the healthy breeders is explained).

In the course of six months to a year, less care need be taken in selecting the earthworms to be fed the frogs. By that time, the earthworm bed should be full of annelids.

The frog farmer should select a secluded spot in which to feed earthworms to the frogs, for they are sensitive creatures and seldom eat in public. Another frog eccentricity is the fact that it will not eat anything unless it (the food) moves.

One of the best methods for feeding earthworms to frogs is to prepare a metal tray or trays, not more than three inches deep. These should be set into the ground so that the edges are level with the surface of the soil. The sides of the trays should be slanted inward from, the bottom, to prevent the earthworms from crawling out.

When the earthworms are transferred to the trays (with a small amount of soil from the culture bed) the contents of the trays should be thoroughly soaked with water. This is done to force the earthworms to the surface, giving the frogs easy access.

Frogs will devour all the earthworms the frog farmer sees fit to give them. It is up to him, therefore, to determine the amount of earthworms he should feed his frogs. In time, he can adjust his earthworm supply to meet his demands.

The second method of feeding earthworms to frogs will, I believe, eventually displace all other methods now in use.

In this method, the earthworm culture bed is placed within the confines of the frog pond enclosure, near the bank of the water.

In preparing this bed, and before the culture is placed in it, it should be wired for electricity as approved by the new science of electroculture. Between the bed and the electric switch should be placed a transformer, a common electrical device that transforms the electrical current up or down as required.

(Before I explain the use of electricity in this manner, it should be understood that the construction of this electrically-equipped culture bed is probably beyond the ability of the ordinary person whose knowledge of electrically-operated appliances is, at best, elemental. It would be a good idea for the person contemplating the installation of this system to contact his electric maintenance bureau or write his State Agricultural College for information concerning electroculture.)

When this system is properly installed, the culture bed prepared as above noted, the earthworms and egg capsules planted and a sufficient quantity of earthworms matured, the electric contrivance is ready to be brought into use.

The first move is to turn the dial of the transformer down to its lowest point, which permits the minimum of "juice" to pass through the wires in the culture bed. This done, the electric current is turned on. If the amount of current is sufficient, earthworms will appear instantly upon the surface of the culture bed, sent there by the electricity. If the earthworms do not appear, the dial of the transformer should be moved to the next higher notch. If again no earthworms appear, the upward movement of the dial should be continued, one notch at a time, until they do appear.

Once the frog farmer using this system learns the necessary amount of electric energy required to expose the earthworms, he can set the dial at this point for future use.

Experiments with this system have shown it to have two beneficial advantages for frog farmers. First, the larger, mature earthworms, i. e., the breeders, do not come to the surface until a comparatively large volume of electricity is turned into the wires in the culture bed. This assures the breeders comparative safety from capture by the frogs. The second benefit is that the frogs may feed to their hearts' content upon the earthworms that do come to the surface.

In the near future I hope to produce a pamphlet dealing exclusively with the application of electricity to earthworm culture beds. While this is still in an embryonic state, I believe it has many possibilities, and I am working on a few experiments at this writing.

Let us now turn our attention briefly to the bluebottle fly, its larvae, and the manner in which these are procured by the frogs.

Lesson 6 in *Part 2* of this series, *Putting the Bluebottle to Work*, explains the method by which the larvae are developed for a supplement diet for hens.

For the frog farmer, the same general principle prevails. Presently, there is a screen-like basket being used that is recommended by certain frog culturists. Decayed fish or animal by-products are placed upon it as an invitation to the bluebottle fly to "blow."

I can recommend no better device than these baskets. They should be suspended over the earthworm feeding trays, high enough to permit the air to carry away the odor, yet not so high as to discourage the bluebottle flies from "blowing" the decayed matter on them.

It seems to me that this dual system of feeding earthworms and fly larvae to frogs in captivity—both foods being the natural food of frogs—should be highly productive of healthier, bigger and better frogs. This is a goal which all frog farmers hope to someday reach.

> *Relation and connection are not somewhere and*
> *time, but everywhere and always.*
> —Emerson

Housing the Earthworm Stock

::

*Building an earthworm farm — The windowsill
earthworm farmer — Small stock all that is
necessary — Earthworm farming in gallon
cans — Economical way to start — Farming
in vegetable lugs — One of the finest methods
— Proved by many years of experimentations
— Farming in master beds or banks — What
the U. S. Department of Agriculture says about
this type of earthworm breeding*

In this lesson, we approach the vortex of earthworm farming
around which everything tending toward success revolves—
the containers in which the earthworm stock is housed.

These containers are to the earthworm farmer what chicken
coops are to the poultryman. Though, the labor required to keep
them serviceable is far below that necessary for efficient poultry
raising.

For continuity's sake, let us follow, step by step, an earthworm
farmer's progress beginning with the smallest type of container
toward his ultimate goal—the housing facilities for one or more
millions of earthworms.

Our first step in this figurative journey will be with:

The Windowsill Earthworm Farmer

Countless thousands of people take delight in decorating the windowsills of their homes, apartments or rooms with flowers. And a surprisingly large number raise a limited amount of vegetables in windowsill boxes.

Regarding this subject, Dr. Martha B. Carey, of Los Angeles, writes:

> "Earthworms have interested me for several years, but I have had to confine my own practical experience with them to window boxes in an apartment house. I have four boxes—30 inches long, 8-1/2 inches wide and 7 inches deep—in which I grow many varieties of flowers and enough carrots, onions, lettuce and parsley to supply my own table..."

A convenient and efficient method for the windowsill earthworm farmer desiring to develop a stock of earthworms from say a tablespoonful (about fifty) egg capsules may be satisfactorily begun in a flower pot.

An ordinary flower pot, not less than six inches in height, should be filled within an inch of the top with rich soil. A small amount of manure and peat moss, if available, would also be beneficial.

Place the earthworm capsules in this mixture and set the pot in a saucer of water. Less than a teaspoonful of corn meal should be sprinkled over the soil and a few grains of barley added. Both of these are food for the earthworms, though barley has the added advantage of sweetening the soil.

Soil in containers in which earthworms are bred and raised becomes so rich in earthworm castings that the soil must be sweetened. If the earthworm farmer was only going to discard the castings, sweetening the soil would not be necessary. But earthworm castings, being of high nutritional value to plant and vegetable life, are constantly kept and used and reused. This system of using earthworm castings may be best explained by comparing them to a sponge. One may fill a sponge with water, squeeze it out and keep this up almost indefinitely.

(Parenthetically, the writer wishes to refer the reader to the following lesson regarding food and feeding methods, where complete instructions for all types of containers will be found. In this lesson, I deal primarily with the types of containers used and the methods employed to prepare them for soil, food and earthworms.)

Under ordinary conditions, the fifty earthworm eggs will have hatched and begun breeding in about three months.

Care should be taken to keep the saucer well filled with water. The pot should never be watered from the top.

In from three to four months, the windowsill earthworm farmer should be ready for the first harvest of earthworms. At that time, another flower pot should be filled with soil and food, as described in the following lesson.

The new pot is then placed in a saucer of water. The first pot is removed from its saucer and placed on top of the new container. Through the standard hole in the bottom of the pot, the growing and breeding earthworms will pass from the upper to the lower pot, leaving their eggs behind them.

Earthworms breed so rapidly that from this period onward the windowsill farmer will be surprised at the speed with which the earthworm stock increases.

Either earthworms or their eggs (or both) may be transplanted from the breeding pots to flower boxes or other pots containing plants as soon as the farmer desires to do so.

As the earthworm stock increases and more and more flower pots are needed to hold them—care should be taken in observing that they are not too crowded (100 mature earthworms being the maximum)—the windowsill farmer graduates to:

Earthworm Farming in Gallon Cans

These containers are made ready to receive the culture and stock by punching three holes, equidistant, about two inches above the base. Holes are punched in them for the purpose of permitting

113

surplus water to drain off. Were the water to gather in the bottom of the can, it would sour. The resulting chemical action would harm the earthworms housed within it.

When the gallon cans are properly filled with culture, earthworms and their capsules—not to exceed one hundred of the former and three hundred of the latter—are placed in them.

Like all containers for breeding and raising earthworms, gallon cans should not become overcrowded. Under no condition is crowding of earthworms advisable.

Space required for storing this type of container doesn't need to be a problem. Twelve cans may easily be placed on a shelf on the back porch, in the cellar or in the garage.

At no time should more than 600—500 is preferred—growing and breeding earthworms be housed in a one-gallon can.

The gallon can earthworm farmer will find that his stock increases much faster than it did when he was breeding and raising them in flower pots. Now his stock is approximately tripled, and the increase in breeders will be from three to four times what they were during the flower pot stage.

In a few months, the number of gallon cans required to house his constantly increasing family of earthworms will be such that he will be ready to set these aside and turn to larger containers.

He is now fast approaching the stage when he may rightly consider his stock of earthworms enough to encourage him to give considerable thought to markets and marketing. He is now ready to begin:

Earthworm Farming in Vegetable Lugs

Vegetable lugs are both popular and practical for the earthworm farmer. They are easy to handle, weighing less than fifty pounds when properly prepared to receive the earthworm stock. (See following lesson.)

Vegetable lugs are purchasable at any market for from three to five cents each. Their approximate inside measurements are 17 inches long, 14 wide and 6 deep.

A good container to use, weighing not over thirty pounds when properly filled, is the common butter box. Its approximate inside measurement is 10 inches long, 9 wide and 6 deep. This container is the choice of women earthworm farmers because it is compact and easy to handle.

Both the vegetable lugs and butter boxes are prepared in the following manner.

In the bottom of each, six quarter-inch holes should be punched or bored. These should be more or less equidistant, about four inches apart, three on each side of the bottom.

These holes play a dual role. First as drainage for surplus water, and, second, to permit the travel of the earthworms from upper to lower boxes. (See following lesson.)

High in the center of one end, bore a hole large enough to receive the nozzle of an ordinary garden hose. This hole is to facilitate watering the stock when the lugs are stacked.

Butter boxes should not contain more than 800 growing and breeding earthworms. Vegetable lugs may safely house 1000 earthworms without fear of crowding.

When the vegetable lugs become numerous, say 24 to 36, the earthworm farmer needs considerably larger quarters. He now enters:

Earthworm Farming in Master Beds

Master beds (or pits, piles or banks, as they are frequently called) are the ultimate goal of every earthworm farmer. With these—for he may have need for a number of them—he will be in a position within twelve months to supply the demands of his clients.

Let us consider a master bed as suggested by the United States Department of Agriculture in its Farmer's Bulletin, No. 1569, from which we learn:

> "Where it is desired to store or rear earthworms for sale, a larger container placed out of doors is desirable.
>
> For this purpose a tight box, preferably constructed

of tongue-and-groove material, is suitable. It should be at least 18 inches deep and of a size proportionate to the number of worms it is proposed to handle. A box 18 by 36 by 60 inches will serve very well for several hundred large worms. If the exterior of the box is well tarred, it will last much longer in the soil than if untreated. Creosote is not recommended for this purpose because of its possible effect on the worms. In any case, the inside of the box should not be treated with either of these substances, but, it may be waterproofed by painting with hot paraffin wax.

The box should be supplied with a well-fitting lid, which should project sufficiently over its edges to prevent flooding during heavy rains. It should be set into the soil with the upper 2 or 3 inches projecting above the surface, in a fairly well-drained place, and should be shaded to prevent the temperature of the interior from rising too high in midsummer. A temperature of 75 deg F. or higher is quickly fatal to earthworms under most conditions. The box should be nearly filled with good soil which is damp but not wet. The richer the soil is in humus the better, as the worms require less artificial feeding in rich soil than in poor. A loamy soil is preferable, but very sandy soil is not suitable.

After the box has been stocked with worms, the surface of the soil may be covered with a layer of cut sod if desired, but a very excellent covering consists of well-decayed leaves, which form a considerable part of the natural food of earthworms. In dry weather it will be necessary to moisten the soil in the box occasionally, but in doing so care should be taken to avoid flooding it, as too much water is injurious to the worms. Freezing kills earthworms, and in severe climates, where the soil commonly freezes to a depth of a foot or more during the winter, it may be necessary to protect the soil in the box from frost. Winter protection may be secured by giving the box a generous covering of half-decayed manure or compost.

Although under the conditions just described earthworms can live for a long time without artificial aid, it will be found desirable to feed them a little fat occasionally, in the form of chopped beef suet, or a little sugar in some cheap form. One dealer in earthworms claims to have been very successful in feeding worms ordinary molasses spread on the surface of gunny sacking or burlap, which is simply laid upon the soil with the sticky side down and moistened occasionally. The worms undoubtedly will reproduce more rapidly and be more thrifty if they are well fed. When the worms obtain insufficient food they shrink rapidly in size and lose vigor."

One may say, and rightly, that the compost bed mentioned will be the earthworm farmer's most expensive item. There are others which are just as good.

One of these is turning old bath tubs, porcelain or metal-lined, into compost containers. This is best done by sinking the tub, or tubs, within a few inches of their rims. A screen should be placed over the drain hole to permit water to pass out, but to defeat attempts of the earthworms from escaping.

Other pits, beds or banks may be made according to one's own ideas.

At Highland Park, California, two unused wells have been filled with compost and earthworms are now being bred and raised in these.

These pits are colloquially referred to as "banks," principally because from them the farm draws its needed supply of both earthworms and their capsules.

These "banks" are divided into quarters, with lattice work partitions. Each quarter is developed independently of the others, with the results that as one quarter is being drawn upon, the three others are developing. With the depletion of one quarter, fresh compost, food and earthworms and their capsules are placed in it and the farm begins to draw the next quarter, and so on, ad infinitum.

Our figurative journey has taken us through all the popular and approved housing facilities for earthworms bred and raised for private and commercial purposes. When the simple rules and suggestions presented are followed, the earthworm farmer will realize that he can spend more money for equipment, but he cannot extract more or better service from it.

And now, with the knowledge of how to prepare containers for earthworm breeding and raising, we turn our attention to the type of soil and food used in them to keep our friend the earthworm healthy and productive.

> *Those who have finished by making all others think*
> *with them have usually been those who began by*
> *daring to think for themselves.*
> *—Colton*

Lesson 5

General Care
and Feeding of
Earthworms

::

*Sound business advice — What successful
business men must know — How Charles
Darwin worked with his equipment — How he
worked with his records — Earthworm farmer
need not have expensive equipment — But he
should know earthworms — How to acquire
this knowledge — One last tip for beginning
earthworm farmers*

In the preceding lesson we saw how various containers to house earthworms were constructed or procured. Now we shall consider the matter of feeding and caring for the earthworms so that they will live, prosper and multiply according to the natural laws governing them as a specie.

The earthworm farmer should constantly keep in mind that he must attend to his stock just as he would have to attend to a flock of chickens, a drove of milk cows or a kennel of dogs, or any other specie of fowl or animal he was raising for private use or commercial disposition. However, the time and labor required for earthworms, whether they are in the hundreds or the hundred thousands, is far less than the farmer would have to expend on any other animal he was breeding and raising in captivity. This lesson will demonstrate the insignificance of time and labor in tending earthworm stock.

In approaching the subject of earthworm food and feeding, picture an earthworm farmer who is just entering the field.

Before, or within a few days after the arrival of his original stock of earthworms, the fledgling farmer should prepare a compost pit, or, if more convenient, a pile. This compost is definitely essential and should be kept complete at all times. It is, one might accurately say, the soil reservoir from which the earthworm farmer draws almost weekly.

Let us assume that the compost will total three bushels. It should be prepared like this:

Either in the pit or in a pile mix one-third manure, one-third soil and one-third peat moss (or substitute). To this may be added much of the kitchen waste, except acids, citrus rinds or scouring powders. Tin cans may also be added, for, as they rust and disintegrate, they are absorbed by the compost.

To mix the above, the following instructions should be followed. The manure should be either:

1) Karakul sheep
2) horse manure, preferably from grain fed animals
3) chicken droppings
4) cattle
5) rabbit

After many experiments, we have found that Karakul sheep manure is the best of all. We all know that every plant, every weed, and every form of vegetation contains elements, which in time contributes to the upbuilding process of Nature.

Karakul sheep do not seem to be discriminating in the matter of their feed. They eat almost everything that grows and in so doing they acquire all the elements of the various forms of vegetation. Therefore, their manure also contains all these life giving elements. A good grade of soil should be used, preferably a sandy loam. This should be thoroughly screened before it is mixed with the manure and peat moss.

Peat moss has exceptionally fine properties for an earthworm compost pit and for use in various types of earthworm containers, as well as for soil in general. Its chief advantage is that it is highly absorptive, absorbing from ten to fifteen times its own dry weight in water. It is an organic material, brown in color and of spongy consistency.

There are a number of grades of peat moss, the best coming to us from the boglands of Germany. However, any peat mosses minus an alkali content may be satisfactorily used. Many domestic peat mosses are high in alkalines and should be shunned by the earthworm farmer.

The use of peat moss is advisable, principally because it will reduce frequent watering of the earthworm stock. It has little, if any, food value; blends easily with soil and is unequivocally superior to any substitute yet known for use in earthworm culturing.

However, if peat moss is unavailable, wood shavings or sawdust may be used. These may be from all woods except redwood. Redwood shavings and sawdust will kill earthworms!

Peat moss should be well dampened before it is mixed with the compost or used for any purpose by the farmer. This assures easier mixing and diminishes the chances of a sudden gust of wind scattering it. Peat moss is purchasable in bales, the most economical size for the earthworm farmer being one hundred pounds.

Screening is very important. The more often the compost is screened, the better it will be as earthworm food. And not only does the screening mix the various elements, but it has a tendency to break them down—a condition always advantageous to the root zones of plants.

Let us assume that the beginning earthworm farmer's stock came to him in 216 sixteen-ounce spice cans. These are technically known as earthworm spawn bricks. Each can or brick contains approximately one hundred egg capsules and growing earthworms, making a grand (approximate) total of 21,000 egg capsules and growing earthworms.

At his earliest convenience, the earthworm farmer should transfer the contents of the spice cans (the spawn bricks) to vegetable lugs. (See preceding lesson.)

This transfer includes the following operations:

In the bottom of the vegetable lug (prepared as described in the preceding lesson) should be placed one-quarter of a gunnysack (burlap). It should be laid flat to cover most, if not all of the bottom of the container. Upon this should be placed some fresh compost from the pit or pile—to a depth of about two inches. Then empty the contents of eighteen of the spawn bricks into it. Cover this with more compost; scatter a small amount (about a tablespoonful) of corn meal or walnut meal over it. A handful of walnut shells may be added. Now, another quarter of a gunnysack—or half or whole if you are so inclined—should cover the contents of the lug. Dampen this thoroughly, using about two quarts of water. Sprinkle about an ounce of barley seed over the burlap and the lug is now ready to be set aside.

The lug should be placed on a flat, heavy board, metal plate or flat piece of heavy tin. Otherwise, the earthworms will crawl through the holes or openings in the bottom of the container, burrow into the ground and disappear.

The earthworm farmer should always remember that his earthworms must constantly be kept under control.

Corn meal, walnut meal and walnut shells are placed in the lug as food. Barley, likewise, is a food, but it serves the additional purpose of keeping the compost sweet.

When the 216 spawn bricks have been emptied into twelve lugs—eighteen to a lug (prepared as described above)—numbers, from 1 to 12, should be conspicuously painted or attached to each lug. The dozen lugs should then be stacked in numerical order in three rows, four deep, with each bottom lug on a flat, level surface.

Except to assure himself that his earthworm stock is adequately watered, the earthworm farmer does not disturb his lugs again until after three weeks. At this time, the position of the lugs are reversed. It is because of this change in position that it is important to number the lugs.

Earthworms always have a tendency to work from upper to lower cases. Reversing the order assures the earthworm farmer distribution of his stock through the various containers and offsets a chance of crowding in one or more lugs.

As each lug is taken from its original position, the green, sprouted barley should be torn off—and, if convenient, thrown into the compost pit—the gunnysack again dampened with water, a small amount of fresh food added and another ounce of barley sprinkled on top. This lug is then placed so that it becomes the bottom container of its stack. Each lug is similarly treated until all twelve lugs have been reversed.

This reversing of the lugs should be done every week or two. Unlike most other animals raised in captivity, the earthworm does not require perfectly-timed and regular attention.

The earthworm farmer will be ready for his harvest in from ninety to one hundred days after the arrival of his original stock. At this time, all the original egg capsules will have hatched and from forty to sixty percent will have laid one or more eggs.

This increase in the earthworm farmer's stock should be sufficient to allow him to market approximately 8000 egg capsules, growing and, in a small percentage, breeding earthworms. He should not appear too anxious to dispose of his first harvest. It is much sounder business to retain that crop and "put it back in the business."

Here is how this is accomplished:

Twelve new lugs, twelve previously prepared quarters of gunny sacks and handy containers for food—corn meal, walnut meal and walnut shells—should be conveniently placed on a bench of congenial height. On the bench should be a flat board—or the bench top itself will be enough if it is free from holes or cracks—or a metal plate or heavy sheet of tin.

The earthworm farmer begins this work from the top lug of the first row. The sprouted barley is removed and discarded. The burlap bag—which is used to retain moisture and keep out bright light—should be placed to one side. Never attempt to pull any earthworms out of the bag. By doing so could pull them apart, injuring them seriously if not fatally. Left untouched, they will crawl out of their own volition, at which time they may be rescued and returned to the compost.

The contents of the lug is then dumped in the center of the bench. With hands, build it into a pyramidal pile and leave it exposed for fifteen to twenty minutes.

During this period, the earthworms will burrow toward the bottom of the pile, permitting the earthworm farmer to begin his harvest without unnecessarily annoying the breeders.

The egg capsules, with a liberal amount of soil (and additional soil from the compost pit) are now gathered and placed in one of the prepared lugs. The pile should be divided with about one-half of the soil with as many egg capsules as could be found placed in one lug and the balance of the soil containing the breeders placed in the other. Both should be fed and watered as explained.

In transferring the growing and breeding earthworms into the second box, the farmer will quickly learn to recognize culls. These become readily distinguishable following a few practices in caring for the earthworms. Culls are either pale or of unusually large proportions.

Our healthy earthworm, the type described in Part One of this work, is a rich reddish animal, seldom longer than four inches. Both types of culls should be destroyed. The large earthworms are atavistic and are not to be desired on a well managed and well operated earthworm farm.

In harvesting the egg capsules for commercial disposition, the earthworm farmer should have small spice tins containing a small amount of food and compost. Into these the capsules may be dropped the number noted and prepared for shipping.

Summary

::

Important Rules for Beginning Earthworm Farmers

1) Number lugs prominently.

2) Gunny (burlap) sacks, cut in quarters, should be free from alkali or any caustic.

3) Place quarter of sack in bottom of each lug.

4) Place not more than two inches of compost in bottom of lug, on top of quartered sack.

5) Empty spawn bricks into the lug, eighteen to a lug.

6) Feed about one tablespoonful of corn meal, walnut meal. (Cottonseed oil may be used).

7) Place on top of this another one-quarter of gunnysack. (Feeding may also be done by placing the gunnysack on top lug first, sprinkling the food over it).

8) Sprinkle about one ounce of seed barley over top of gunnysack.

9) Contents of lugs should be kept moist at all times, the quantity of water used may be judged by amount of evaporation.

10) To facilitate harvesting egg capsules and earthworms, contents of lugs should be allowed to dry out.

Care and Harvesting

1) For best results, lugs should be reversed every two weeks.

2) When reversing lugs, remove barley growth, re-feed and re-barley with fresh seed.

3) Lugs should be reversed every two weeks thereafter.

4) When reversing is completed, water thoroughly, using about two quarts of water for each lug.

5) It is not advisable to fill lugs higher than three-fourths of their depth. This is suggested to reduce the weight. A full lug will weigh in the neighborhood of fifty pounds. Lowering the contents will reduce this weight to as low as thirty pounds.

6) In transferring breeding stock from one lug to another, the top sack of the first lug may be used as the bottom carpet for the new lug. Bottom sacks remain serviceable for about three weeks.

A potential commercial field which was not discussed in the body of this work is that of the extraction of oil from earthworms for medicinal use.

I have repeatedly experimented with earthworm oil and have found it has exceptionally fine penetrating proclivities.

I purposely refrained from mentioning this subject in the lesson devoted to markets for earthworms, principally because the experiments I have so far carried on (in co-operation with a number of chemists) have not progressed sufficiently to warrant a lengthy discourse on the subject.

It is an accepted fact that earthworm oil has medicinal value. But to date, myself and associates have not found a satisfactory method for extracting the oil in sufficient quantities to meet even a small commercial demand. I have used very primitive methods to extract the oil, which must be done by heat—to which earthworms are very susceptible. The method used was slow, tedious and annoying, though I have managed, by painstaking efforts, to garner well over a gallon over months of experimenting.

At the present writing, experiments are being prepared which will employ electric heat.

On the whole, I feel confident that, in time, a suitable and wholly satisfactory method of extracting earthworm oil in sufficient quantities to warrant its consideration in commerce will be available. When that time comes, which I hope will not be in the too distant future, I shall make it a point to advise all interested readers of this book. Should the reader desire further information on this subject of earthworm oil, I suggest he contact Dr. O. M. Crause, 4575 Melrose Avenue, Hollywood, California, in care of The Associated Laboratories.

Another subject purposely left out of the preceding parts is that of kelp ore. This ore is, geologists say, the result of a cataclysmic earth disturbance that threw countless tons of kelp (seaweed) out of the sea. In time, this kelp became buried under many tons of earth, ossified and remained undiscovered for many eons.

Seaweed—kelp is a Saxon word meaning charred seaweed—is one of the oldest plants known. It antedates by millions of years the first forms of plant life on land—the mosses. It is potent in minerals, contains 14 of the 16 elements in the human body, and is especially well impregnated with iodine, iron and phosphorus.

Kelp is recognized today by progressive physicians as a highly nutritional food supplement for man, beast and fowl.

Of the 82 known elements, 35 are found in the sea, and of these 35, kelp contains 27.

I have been experimenting with kelp ore for over a decade and use it, in solution, myself and recommend it to my friends.

Using kelp ore in conjunction with earthworm culture is advisable. But, due to its many recommendable qualities, due to the fact that much of its potency can be taken off in solution for human consumption and the residue used for earthworms, I must regretfully refrain from going into its use by earthworm farmers at this time.

Considerable detail is necessary to clearly explain the use of kelp ore, the benefits derived and why. To cover this, I am preparing a small Paper on the subject. Persons writing the author may receive this information.

The greatest events of an age are its best thoughts. It is the nature of the thought to find its way into action.
—Bovee

Conclusion

⠃

The purpose of these final words regarding our friend, the earthworm, is to draw together a number of loose ends that were inadvertently left out of the main body of this work. In addition, the writer wishes to explain briefly the physiology of plants. Particularly that dealing with the roots of plants, for it is in the root zone that the earthworm plays its major role.

In *Lesson One, Part One*, we saw how all life began on this planet in the waters. Only superficially was it mentioned that the origin of plants was also in the water.

First of all, let us differentiate between animals and plants. Offhand, the average person would say that animals are sensitive and mobile; that plants are not sensitive and are rooted or stationary.

Such a definition is satisfactory, though it is not correct. Among the zoophytes (Greek; zoo—animal, phyte—plant) there are immobile animals and sensitive plants. These latter respond to vibrations of the air around them and to the touch of a foreign body. In addition, there are marine animals that remain stationary from birth to death, and marine plants that travel extensively.

So we see that neither movement nor sensitiveness are essential distinctions when we are pinned down and asked to explain the difference between animals and plants. However, men of science have an infallible method of classifying them, and that is by the manner in which they acquire nourishment. Animals "eat"; plants "absorb" food.

Generally speaking, plants have two sets of "mouths"—their roots and leaves, or, in the absence of leaves, stalks or stems.

We are especially interested in the physiology of roots because of our friend, the earthworm. It is important to spend a little time on roots so that we shall all have a clearer picture, a sound foundation, from which to visualize the work of the earthworm in the root zone.

There are more varieties of roots than there are plants, and there are many thousand species of plants. There are roots that store water for the plant (many desert species); roots that store food (potatoes, onions, etc.); roots that support plants and plants that support roots, like the banyan tree. There are roots that are parasites, some of which never need contact with the earth. There are others, like the wild fig tree—which is a relative of the banyan—that comes to life high on the trunk of the eucalyptus. As the fig tree grows, it sends shoots of new roots earthward. Here they burrow into the soil, eventually supporting the tree itself.

The earth's first plant life was in the form of moss. It was in this age that roots originated! Gradually, through hundreds of thousands of years, evolution developed other and sturdier-rooted mosses. Then, during the Devion Period—a geological era in the Paleozoic Age—ferns, clubmosses, horsetails and trees so large that they would make California redwoods look like celery stalks in comparison, grew and prospered for countless centuries.

From the very beginning of plant life on land, roots have performed a dual role—as support and as a feeding apparatus for the plant.

The fundamental purpose of roots is to gather water and certain salts from the soil. Through the chemistry of the plant's system, this becomes nourishment.

The outstanding peculiarity of roots is not observed without close investigation. This peculiarity is to be found in the myriad of hairlike, fibrous fingers through which the nourishment for the plant is sucked up. The brush or "tap roots" common to many plants are almost exclusively engaged in anchoring the plant in the soil.

The small, hairlike, inconspicuous roots (which may be observed by washing the root area carefully) are the real "mouths" of the plant. Nourishment is not actually sucked up by these hairlike roots. The expression is used merely in a figurative sense. The action is entirely chemical.

Water is attracted by the plant roots by the sugary content of the minute cells of which the roots are composed.

Plants take nitrogen, oxygen and hydrogen from the soil. Carbon, another highly essential element needed by plants, is received through the leaves and (or) stalks.

The leaves or stalks compose the second set of the plant's "mouths." These are filled with a substance of very peculiar properties. This substance was named chlorophyl by the Greeks and means "green leaf."

Chlorophyl is an ingredient in the cells of leaves, stems and stalks. It is the chemical that breaks up the carbonic acid gas in the air which the leaves "breathe." From this breaking up process, carbon is extracted, mixed with hydrogen. The oxygen is set free, for enough oxygen is taken in by the roots to satisfy the plant's needs.

Chlorophyl is the substance that makes plants green. It is developed by the plant as needed, but it cannot be developed without sunlight, or a satisfactory, albeit inferior, substitute. Take light from plants and chlorophyl disappears. Example: when soil is piled around the stalks of celery just above the ground, the portion covered becomes white. Lack of light has forced the chlorophyl up the stalks.

Sunlight and chlorophyl are as important to plants above the soil as roots are under it. And it is because of their importance that plant life, from the tenderest orchid to the sturdiest oak tree, spread their leaves and branches about so haphazardly. Each is striving to absorb as much sunlight as it possibly can.

Not only do the leaves of plants "breathe" moisture through their pores; they must also prevent the plant from losing too much water through them. But I fear I am getting too far away from our real interest, roots. Persons interested in this subject can consult their local librarian for suitable books on this especial subject.

We have already seen how the earthworm, by its constant eating, pulverizing what it eats and excreting it as castings, unconsciously, but with extreme efficiency, prepares the soil. And this to a degree that its mineral and chemical qualities are more easily absorbed by the tender roots of plants.

This pulverization into a fine powder of every minute morsel the earthworm swallows, ultimately results in:

1) a healthier plant
2) a plant richer in chlorophyl
3) more fertile, healthier seeds
4) rapid, even growth
5) if edible for man or beast, a plant richer in food elements.

These advantages are the natural outgrowth of the burrowing earthworm, substantiated by men of science, and therefore not mere words to befuddle or annoy the reader.

In *Soil Science*, October 1935, two members (W L. Powers & W. B. Bollen) of the Oregon Agricultural Experimental Station, wrote:

> "... Earthworm castings were found in the fir forest under the litter in the crumbmull on top of the mineral soil. The castings were collected from the surface on cut over land where there was a little litter... The work of the earthworms appears to have little effect on reaction. There is evidence of a build-up in base exchange capacity, *and the nitrogen and organic matter are much higher in the casting than in the parent soil.*" (The italics are mine.)

In an accompanying table the percentage of organic matter and nitrogen were as follows.

	Maple and grass litter	Soil	Casting
Organic matter	73.14	6.52	34.66
Total nitrogen		0.228	0.672

Clarence Burnham, research fellow in soils, assisted in the determinations.

These investigators found that nitrogen was much higher in the earthworm castings than in the native soil. This fact is a point all persons interested in plant culture will find extremely enlightening.

Nitrogen is the first fertilizing principle to become depleted. But with earthworms functioning in the soil, nitrogen content is increased and will continue to remain as long as there are enough earthworms burrowing in soil containing traces of it. And not only will the earthworms triturate it and make it available for the plant roots, but it will quite probably be brought to the root zone by passing through the alimentary canal of many earthworms.

Let us shift our line of thought for a few minutes.

The new word incorporated in this work for the first time—*Soilution*—may be misconstrued by casual readers as pertaining to the so-called new water culture method. This method, in addition to being currently known under various and attractive trade names, is also called, chemical culture, hydroponics, tank farming, tray agriculture and what-not.

Hydroponics has been sweeping the nation lately under the guise of being "new." It is not new—by seventy-five years.

D. R. Hoagland of the University of California in the February 1938 issue of *The Pacific Rural Press*, writes:

> "First of all, it should be recalled that plants have been grown in solutions containing the essential nutrient salts (plants, of course, will not grow in pure water, so the term water culture is not accurate), by hundreds of investigators during the past three quarters of a century. Their purpose in growing

plants in this way has been to study the laws of plant growth, under controlled conditions. During the same period, another method of artificial culture has also been extensively employed by investigators, that of growing plants in silica sand irrigated with nutrient salt solutions."

Hoagland is not at all optimistic about this "new" form of plant raising. His few favorable comments are offset with

"however, without expert knowledge of the water culture technique, commercial success is unlikely."

And again,

"Contrary to some statements, plants are not protected against diseases (except soil diseases) or insect pests by growth in water culture. Also, it has not been proved that food produced by water culture has, in general, nutritional value superior to that of feed produced by soil, with respect to mineral content of the food."

Hoagland also tells us that

"most amateurs are not in a position to make mixtures of nutrient salts for themselves, and various firms and individuals offer for sale small packages of salts ready to dissolve. Some of the prices are from twenty to fifty times the original cost of salts."

I believe the following letter is worthy of the reader's attention, therefore I include it here:

"...the subject (of earthworms), instead of being of minor importance, is one of the greatest of modern times. Words are inadequate to express the real importance of this lowly creature to the life and pleasure of man. The more one studies the matter, the more vast become the possibilities. In June of this year, I will move to a small place I have at Barwick, Georgia. There I hope to experiment further, as there are eighty pecan trees, ten years old, on the place that I want to treat with castings and earthworms to see what can be accomplished in a backward grove.

This morning I measured petunia plants with branches twelve and one-half inches long, leaves three and one-quarter by two and one-quarter inches. The plants are growing in a box that has been full of earthworms since last spring, working in manured soil.

These plants were from a bed that had been enriched both last year and this year with various fertilizers, in which companion plants of the same age and specie have branches only three and one-half inches long and leaves only one and one-half by one and one-quarter inches.

I have been showing these to some ladies who have flowers, as I feared they would not believe me if I told them without (showing them) the living proof. In fact, it almost looks impossible, but I know the particulars. I wish I were capable of taking a good photograph of these plants while they show such a vast difference in size and thrift.

Perhaps we can work out a plan which I can put into operation when I move to Barwick. It should be very interesting, for there are a lot of rich estates in and around Thomasville, eighteen miles from my future home, such as those of the late Hon. Robert Bingham, one-time ambassador to England; Mrs. Howard Payne Whitney, Ruth Hanna McCormick and others, who might become interested in the use of earthworms for their landscape projects.

At Albany, Georgia, not far from Barwick, there are many more such estates and great game preserves, consisting of as many as 20,000 to 30,000 acres, owned by wealthy people from the north...

I have made a living with hens, and feel that with the information in Part II of *Our Friend the Earthworm*, I can make more than I have in the past. As to housing costs, that is a low item in this section—less than twenty-five cents per hen under usual conditions—but your feeding plan is a lot cheaper and more in the way Nature intended...

I see no way of improving your books, unless it

would be that you impress your readers with the fact that your writing is very condensed, and for them to read the book over and over; or rewrite them in more extended form, for the subject is broad and far deeper than casual notice makes it appear—as broad as every square foot of tillable soil on the surface of the earth.

I am fifty years of age, far too old to try all the experiments I would like with them before my allotted time shall end."

R. A. Caldwell, Faceville, Georgia

Thought means life, since those who do not think do not live in any high or real sense. Thinking makes the man.
—A. B. Alcott

Appendices

My Grandfather's Earthworm Farm

by George Sheffield Oliver

::

*The story of a self-contained farm of 160 acres,
maintained in ever-increasing fertility over a period
of more than sixty years, through the utilization
of earthworms. A true story related to Thomas J.
Barrett by Dr. George Sheffield Oliver.*

When, as a small boy, I went to live with my grandfather, George Sheffield, in northern Ohio, I found him living on a model farm of 160 acres, which he had farmed continuously for more than sixty years. He was a. man who loved the soil and took pride in every detail of his farm. I remember him as a tall, striking figure, of the type of Edwin Markham. In fact, in later years, when I came across a picture of the poet Markham, I was struck by the close resemblance of the two men—their features were almost identical and they could have easily been taken for twins.

Some of my most pleasant memories from the time I spent on this farm are the daily horseback rides I took with my grandfather. After all these years, I can still see him, at the age of seventy-five, riding with the ease and grace of the practised horseman, swinging into the saddle with the facility of a man in his prime. At that age, he still took delight in riding the young three-year-olds. He lived to the ripe old age of ninety-three.

Originally, this farm-holding had been 1,800 acres, but it had been sold off in forty-acre tracts to former tenants until there remained only the farmstead of 160 acres. It had been my grandfather's practice to select young single men as farm help. As these men reached maturity and married and wanted to establish homes of their own, my grandfather would set each of them up on a tract of forty acres or more, assist them in getting started, and accept a payment contract over a period of forty years.

And so, his close neighbours were men who, like himself, loved the soil and could co-operate in all community work. My grandfather often remarked that he was making more profit from his remaining 160 acres than he ever made on the original 1,800 acres, due to his lifetime experience, improved methods, and the intensive utilization of earthworms.

The homestead was located at the centre the farm. Four acres of orchard and garden furnished an abundance of fruits and vegetables year round. Root cellars, vegetable banks, canned and dried fruits and vegetables provided for the winter months. The house and orchard were backed by forty acres of timbered land—maple, hickory, black walnut, burr oak, and many other trees native to Ohio.

Incidentally, the farm was fenced with black walnut rails—beautiful timber which would be almost priceless at this time. My grandfather called this timbered tract his park. It was, indeed, a wonderful park, abounding in small game and bird life to delight the soul of a small boy with his first gun. The park was well watered with living springs and a quite generous-sized creek ran through it, large enough to furnish all the fish the family needed. I was designated as the official fish-catcher, a task which I dearly loved.

It is important to get a picture of the layout of the farm, in order to understand its efficient operation without waste of time and energy. It was divided into four tracts of forty acres each. The homestead, with orchard, garden and park occupied one forty. Near the centre of the 160 acres was located the great barnyard of about two acres, with broad swinging gates in each of the four sides, opening into lanes which led into each of the forty-acre tracts. The stock could be herded into any part of the farm, simply by opening the proper gate and driving them through the lane into the particular section that was to be pastured.

Located in the four corners of the barnyard were the strawstacks—alternating wheat stack, oat stack, wheat stack, oat stack. These stacks occupied permanent raised platforms, about six feet above the ground, resting on sturdy walnut posts and covered by small logs, or poles, cut from the woods. The stock had good shelter under these platforms in the winter, feeding on the straw overhead through the cracks between the logs. Plenty of straw was always thrown down for bedding. My grandfather claimed that each kind of straw added valuable elements of fertility to his compost, and he alternated the straw stacks so that the wheat and oat straw would be evenly mixed.

In the centre of the barnyard was the compost pit, which, in the light of my present knowledge, I now know to have been the most perfect and scientific fertilizer production unit I have ever known. This pit was fifty feet wide and one hundred feet long and had been excavated to a depth of about two feet. At each end, evenly spaced from side to side and about twenty feet from the end, a heavy log post was deeply anchored.

These posts were probably twelve to fifteen feet high, with an overhead cable anchored to the top of each post and running to the barn. On these cables were large travelling dump baskets, in which the manure from the barn was transported to the compost pit and dumped each morning. It was then to be evenly spread in a uniform layer. By means of the posts in each end, the manure could be dumped at a spot most convenient for proper handling. With this arrangement of overhead trolley from barn to compost pit, it was possible to clear the barn quickly each morning of the night's droppings and spread the material in the pit without any loss of the valuable elements of fresh manure. This is an important point in the utilization of earthworms for general farming.

Just outside the barnyard ran the creek, which found its source in a big spring in the park. From this creek an abundance of water was piped by gravity into the watering troughs for the stock in barn and yard. Also a flume, with a controlled intake, led to the compost pit, so that when necessary the compost could be well soaked in a few minutes. The homestead occupied ground on a higher level than the barnyard, so that drainage was always away from the house and there was no chance of pollution from the teeming life of the barnyard.

To one side of the barnyard and at a higher level than the floor of the yard was located the ice pond. This pond was so arranged that it could be filled from a flume, leading by gravity from the creek at one end, while at the lower end a spillway was provided so that the pond could be drained. At the proper season, the ice pond would be filled and when the ice formed to the right thickness the annual harvest of ice was cut and stored in the ice house, to provide an abundance of ice for all purposes the year round. The bottom of this pond was formed of a fine-textured red clay. Each spring the pond was drained and with teams of scrapers, many tons of this clay were scraped out and diked around the borders of the pond to weather for use on the compost heap.

And now enters the earthworm. For more than sixty years, these 160 acres had been farmed without a single crop failure. My grandfather was known far and wide for the unequalled excellence of his corn and other grain, and a large part of his surplus was disposed of at top prices for seed purposes.

The farm combined general farming and stock raising; my grandfather's hobby, for pleasure and profit, was the breeding and training of fine saddle horses and matched Hambletonian teams. He maintained a herd of about fifty horses, including stud, brood mares, and colts in all stages of development. In addition to horses, he had cattle, sheep, hogs, and a variety of fowl, including a flock of about five hundred chickens which had the run of the barnyard, with a flock of ducks.

Usually about three hundred head of stock were wintered. The hired help consisted of three or four men, according to the season, with additional help at rush seasons. This establishment was maintained in prosperity and plenty, and my grandfather attributed his unvarying success as a farmer to his utilization of earthworms in maintaining and rebuilding the fertility of the soil in an unbroken cycle. The heart of the farming technique was the compost pit.

As previously mentioned, the pit was fifty by one hundred feet, excavated to a depth of two feet, and it was especially designed to provide a great breeding bed for earthworms. Literally millions of earthworms inhabited the pit and compost heap. Each morning the barn was cleaned, the droppings for the previous twenty-four hours were transported to the heap by the dump baskets on the overhead trolley, and evenly spread over the surface.

The building of the compost heap was an invariable daily routine of the farm work. A flock of chickens were always scratching and working in the barnyard, assisted by the ducks, gleaning every bit of undigested grain that found its way into the manure. While incidentally adding about twenty tons of droppings per year to the material which eventually found its way into the compost heap.

The cattle and sheep grazed around the four straw stacks and bedded under the shelter of the stacks, adding their droppings to the surface and treading them into the bedding material. From time to time the entire barnyard was raked and scraped, the combined manure and litter being harrowed to the compost heap and distributed in an even layer over the entire surface. As the compost reached a depth of twelve to fourteen inches, several tons of the red clay from the border of the ice pond would be hauled in and spread in an even layer over the surface of the compost. The variety of animal manures from horses, cattle, sheep, pigs, and fowl alternated in the heap with layers of the fine-textured clay, rich in mineral elements.

Meantime, beneath the surface, the earthworms multiplied

in untold millions, gorging ceaselessly upon the manures and decomposing vegetable matter, as well as the mineral clay soil, and depositing their excreta in the form of castings—a completely broken down, deodorized soil, rich in all the elements of plant life. From time to time, as necessary (the necessity being determined by careful inspection on the part of my grandfather), the compost would be watered through the flume leading from the creek. This provided the moisture needed to permit the earthworms to function to the greatest advantage in their life-work of converting compost to humus.

Within a few months, the earthworms had completed their work. When spring arrived, the season of the annual plowing, the top layer of the heap would be stripped back, revealing the perfect work of the worms. What had originally been an ill-smelling mixture of manure, urine, and litter, was now a dark, fertile, crumbly soil, with the odour of fresh-turned earth. This material was not handled with forks, but with shovels. There were no dense cakes of burned, half-decomposed manure. My grandfather would take a handful of the material and smell it before pronouncing it ready for the fields. The 'smell test' was a sure way of judging the quality. When perfect transformation had taken place, all odour of manure had disappeared and the material had the clean smell of new earth.

At this time of the year, the beginning of the spring plowing, the compost heap was almost a solid mass of earthworms and every shovel of material would contain scores of them. As I now know from years of study and experiment, every cubic foot of this material contained hundreds and hundreds of earthworm egg-capsules, each of which, within two or three weeks after burial in the fields, would hatch out from two or three to as many as twenty worms.

The newly hatched earthworms became the permanent population of the soil. They were content in following their life-work of digesting the organic material, mixing and combining it with much earth in the process, and depositing it in and on the surface as castings. It was the creation of a finely conditioned, homogenized soil, rich in the stored and available elements of plant food in water-soluble form.

When the spring plowing began, the following method was adopted: Several teams were used with the plows, while two or three farm wagons with deep beds were employed in hauling the crumbly end-product of the earthworms from the compost pit to the fields. The wagons worked ahead of the plows, the material being spread generously on the surface and quickly plowed under.

Seldom was any material exposed on the surface more than a few minutes ahead of the plows. Part of the technique followed was to plow the egg-capsules and live earthworms under, so that as many of the earthworms would survive as possible to continue their valuable work in the soil. Also, it was necessary to plow the worms and capsules under as quickly as possible to escape the voracious, marauding crows which swarmed in great flocks to the feast of worms and capsules so thoughtfully spread for them.

At this time, to my great delight, I was appointed crow hunter. Armed with a light shotgun, I industriously banged away at the crows to my heart's content, killing some of them and keeping hundreds of them at a distance until the plows could turn the earth and bury the worms and capsules safe from the birds and the sun. I estimate that several tons per acre of this highly potent fertilizer material were annually plowed into the fields in preparation for the crops to follow. On account of this technique, the earth was continually occupied by a huge worm population the year round. Annually, a generous 'seeding' with live earthworms and capsules was planted to replenish and help renew the fertility of the earth.

More than forty years after my experience on my grandfather's farm, studies of the earthworms in the soil of Ohio were made by the Ohio State University. In plots of soil covered with bluegrass, on the Ohio State University Farm, they found earthworms in numbers of one million or more per acre. From my experience of almost a lifetime of study and experimentation with earthworms,

I am sure that the earthworm population of my grandfather's farm far exceeded one million to the acre.

In the annual distribution of the fertilizer, my grandfather never completely stripped the compost pit. One year he would begin the hauling at one end of the pit, stripping back the top layers of material which had not been broken down, leaving a generous portion at the other end of the pit as breeding and culture ground. After the hauling of the fertilizer was completed, the entire remaining contents of the pit were evenly spread over the entire surface for 'mother substance' and the new compost heap was begun.

By this method there was always left a very large number of breeding earthworms, with vast numbers of egg capsules, to repopulate the compost pit and carry on the highly important work of providing fertilizer for the coming year. In this warm, highly favorable environment, the worms multiplied quickly.

In my experiments in later years, I determined that certain breeds of earthworms, in a favorable environment and with an abundance of food material to work on, will work ceaselessly in concentrations of more than 50,000 to the cubic yard. Also, that 50,000 earthworms working in this way will completely transform one cubic yard of material per month. And so, in nature we have a constructive force which creates humus with amazing rapidity when given the opportunity. Under proper control, it furnishes a method for utilizing every possible end-product of biological activity through the very simple process of composting with earthworms.

Going back to my grandfather's farm, his regular rotation of crops was corn, wheat, oats, timothy, and clover hay, in a three-year cycle. One forty-acre tract was planted to timothy and clover each year. A crop of hay was harvested and stored for the winter, the field was used for grazing, and finally a crop was turned under for green manure. In this manner, each year one 'forty' was left undisturbed by the plow for a number of months, allowing the earthworm population to work and multiply to the maximum, while converting the organic content of the earth into the finest form of humus. When the clover fields were plowed under, an almost unbelievable number of earthworms was revealed as the sod was turned.

One fact I failed to mention was that this land was not usually considered the finest to begin with. It was a thin topsoil, only six to eight inches in depth over much of the farm, underlaid by limestone. On account of the shallow depth of the soil, deep subsoil plowing was not possible.

I remember how the plows would scoot along on top of the almost surface limestone layer. However, the vast earthworm population penetrated deeply into the subsoil and constantly brought up parent mineral material to combine with the surface soil, which made up for the lack of deep soil. My grandfather often remarked that in all his sixty years of farming he had never had a crop failure. His corn was the finest in all the country and was eagerly sought for seed. He also originated a sweet corn, of a delicious flavour, which was very highly esteemed throughout that section and was known at that time as 'Sheffield corn'. The ears were very uniform and evenly filled to the end, and I remember that the cob of this special corn was hardly larger than a carpenter's lead pencil. My grandfather never sold this corn, but reserved it to give to friends who came from far and wide for the prized seed.

Now, looking back through the long vista of years to the method practised on my grandfather's farm, in the light of my own experience as well as the experience of a host of others, I am struck by the reflection that here was a simple farmer, working without any specialized knowledge of earthworms to begin with, long before Charles Darwin's famous book on *The Formation of Vegetable Mould* appeared. And yet, in an intensely practical way, utilizing all that Darwin later revealed in his great book, but with the exception that Darwin never suggested the 'harnessing of the earthworm' for intensive human use. Darwin's classic study only emphasized the importance of the work of the earthworm in nature, with no practical application to the personal agricultural problems of man.

Before ending this narrative of my grandfather's earthworm farm, I must mention the orchard, the garden, and the fence rows.

The fence rows throughout the farm were planted with a great variety of fruit trees, which were allowed to develop from seedlings. In particular, I remember the cherry trees, some of them fifty feet high and each tree bearing a different kind of fruit. In the four acres of orchard and garden surrounding the house there was produced a great variety of fruit, furnishing an abundance, in season, for the family as well as for many of the neighbours.

In those days, the fruit was not sold. I remember an often-repeated remark of my grandfather upon the care of trees, especially fruit trees. He said, 'Never disturb the soil under a tree. The earthworm is the best plow for a tree and I do not want them disturbed.' The vegetable garden was especially fine, kept wonderfully enriched from the compost pit, the soil being literally alive with earthworms. Flowers, both potted and otherwise, as well as a wealth of shrubbery, beautified the place. For choice flowers, we would use a rich mixture of fine soil and material from the compost pit.

My grandfather's earthworm farm furnishes an example of the technique for utilizing the earthworm in general farming operations, either on a large or small scale. From my observations as a small boy, supplemented by much friendly and loving instruction from my grandfather on the subject of earthworms, and from more than forty years' experience in my own work, I am fully convinced that the harnessing of the earthworm will be one of the major factors in the eventual salvation of the soil. I know that the soil can be made to produce several times as much as the present average, through the utilization of the earthworm.

Introduction to
Harnessing the Earthworm

by Eve Balfour

⬛

'When the question is asked, "Can I build top-soil?" the answer is "Yes", and when the first question is followed by a second question, 'How can I do it?" the answer is "Feed earthworms".'

That is the last sentence of this book. It seems to me particularly appropriate to use it as the first sentence of my introduction, because it serves equally well as a preface to, or a summary of, what the book is all about. This fact symbolizes that in the Wheel of Life, or the Nutrition Cycle, or by whatever other name you prefer to call it, there is neither beginning nor end, but only continuity; an unbroken progression of birth, growth, reproduction, decline, death, decay, rebirth. A continuous flow of substances passing from one form of life to another, round and round the cycle without end.

Dr. Barrett puts it more simply; he says,

> "Earthworms are soil builders, everything else—plant, animal, man and bacteria are food for earthworms whose function is to mix living matter with mineral particles and send them forth on their round once again."

That is the cycle as established by nature and operated by the creative, i.e. living, forces. It worked on an ever-ascending spiral, accumulating richer and more varied life forms, until man arrived upon the scene. It has been left to this supposedly most intelligent of all creation's species to put the wheel into reverse by abandoning creative motive power in favor of consumptive power, i.e., the destructive forces. In doing this, man has enacted the role of the parasite whose ravages destroy the host upon which it is dependent for sustenance.

He has been guilty of this behaviour since the early dawn of 'civilization'. His host is the fertile topsoil forming the surface covering of the globe; a thin covering now, very threadbare in places. 'The wasting basis of civilization', is how Sir John Boyd Orr defines soil fertility. It is man who is responsible for this wasting. Of the fertile cultivatable area of the U.S.A., as it was found by the pioneers, one-quarter is gone forever, so their soil experts tell us, and many million acres are still disappearing annually. The same story comes from South Africa. There are deserts extending for hundreds of square miles which were producing good crops only thirty years ago.

Australia and New Zealand have the same sorry record of man's rapacious exploitation to relate, and even European soil shows signs of the same decline.

The phenomenon is not new. In the name of economic necessity, God forgive him, mankind has destroyed the source of his food since before the days when part of the Sahara Desert was known as the granary of Rome. The two new factors are the speed with which modern man can turn fertile land into desert, and the fact that there no longer exist any new virgin lands for him to discover and exploit. He has reached the last barrier. At last he must learn the bitter lesson of his past mistakes or perish from the face of the earth, like other species, now extinct, who failed to solve the problem of how to co-operate with their environment.

That is the major crisis facing the human race today. It is a challenge beside which, as a recent writer put it, the bickerings of

150

Foreign Ministers sound like the jabbering on Monkey Hill.

Those of us who believe that the living, creative forces are the only ones that can promote and sustain life, know that soil fertility can still be maintained by obeying nature's law of return—vitality in soil, plant, animal and man results. But the time is short, and mass action is required now if it is not too late. The warning has been cried aloud from the housetops by men of knowledge and of the highest repute. From every continent, almost every country, their warning and call to action comes—'The Human Race faces mass starvation—Act Now or your children's children will die.'

Does anyone pay the slightest attention? Very few. Does anyone ever pay the slightest attention to prophets of woe? They persecute the prophets sometimes, but that is about all. The prophets were so frequently right that I have often marvelled at the persistent deafness of mankind to all warnings of preventable horrors to come. I have come to the conclusion that the explanation of this is twofold.

First, in the case of the powers that be; those in authority are always preoccupied by the immediate problems of the moment. They have become permanently myopic and are literally incapable of taking any but the shortest of short-term views.

At the present time, for example, the need for timber, for fuel and housing now, is of such apparently prime importance that it seems to justify the risk that a new desert will result tomorrow. It is a mistaken view and it is the attitude of mind that has produced the dustbowls of the world and landed us in the mess we are in; a mess which makes it increasingly more difficult to opt for the long-term view.

One can have nothing but sympathy with those who may have to choose between the death of hundreds now or millions tomorrow. There is always the chance that if one saves the hundreds now one will not live to see the million perish. So it is—to take a topical example—that the present danger from the atomic bomb appears much greater than that from soil erosion. It isn't. In point of fact it is a mere flea bite to it when considered in terms of the probable survival of the species. But that is the way it *looks*, so those in authority perpetually confuse priorities and postpone action, while the rest of the populations resorts to apathy. I think, from the feeling that nothing they personally can do about it can possibly affect the situation as a whole, 'so what's the use?'

I urge anyone who feels that way about it to read this book, and here I must make a confession. Like other people who have had practical experience of the results of compost-making and organic cultivation generally, I have for long been convinced that in the cycle of life the members of the soil population play a vital part, and that when relying on the return of all possible organic wastes to the soil as our sole method of fertilizing, we are not feeding our crops direct, but *through* the soil population. We have, in fact, a slogan—'Feed the soil population and they will feed your crop.'

Among this vast population, I have always recognized that the earthworm was a creator of soil fertility without equal. This view is confirmed and strengthened by the recent research work carried out at the Connecticut Agricultural Research Station which is reprinted in this book and of which I was already aware. I am even myself a breeder of domesticated earthworms in a small way. For all these reasons, I did not expect to find anything particularly startling or new to me personally, in a book called *Harnessing the Earthworm*. I was wrong.

I did not know, for example, that in fertile soil the weight of bacteria alone amounts to 7,000 lb. per acre. I did not know that anywhere in cultivated soil, however fertile, the natural earthworm population reached two million per acre (Nile Valley). I did not know that in any part of the world, even where intensive propagation of earthworms for soil building was carried out, there were farms where as many as three million earthworms per acre have been recorded. And that populations of between one and two million are quite common.

I did not know that one million earthworms weigh a ton, or that in the course of twenty-four hours each worm will pass through its body its own weight of soil.

Earthworm castings are composed largely of colloidal soluble humus, and are far richer in available plant foods than the surrounding soil. This represents a staggering annual deposit of natural plant fertilizer, quite apart from the continual addition of the dead bodies of the worms themselves as they fulfill their own life cycle.

The figures given in this book of the differences in crop yields obtained from soils of equal fertility, with and without added earthworms, are startling, but not unbelievable once the data given is studied.

While I doubt whether quite such spectacular results could ever be obtained in our climate, earthworms can and do exist in a very wide range of latitudes. Where they can exist they can be increased, and there is no better or quicker way of increasing them than by intensive propagation of egg capsules in special breeding boxes. When transferred as eggs to their final location in garden, field or orchard, they will survive, whereas the mature or growing worm may not.

The technique for this intensive propagation is simple, and Dr. Barrett gives such clear and concise instructions that anyone—whether he starts with purchased stock or native brandlings—can test the claims made for it for himself.

For the English reader, however, there is one serious omission. The optimum temperature for maximum production of egg capsules is given as 70 degrees. While plenty of advice is given for protecting culture boxes or master beds from too great heat, nothing whatever is said about how to protect them from cold.

My own experience may therefore be helpful.

Culture *in boxes* must either be discontinued during the winter, or take place in cellars or heated greenhouses. Culture in master beds can continue, provided these are sunk in the ground and covered up with straw in frosty weather. Attention of course must be given to drainage from below, and prevention of flooding from above.

I consider it highly important that experiments on the lines suggested in this book should be carried out without delay. As organic cultivators are well aware, the principal argument used against them by soil scientists, is based on mathematics. 'The crop takes out more than the compost puts back. The result must be a deficiency.' The organic cultivator replies that the proof of the pudding is in the eating, which he can demonstrate. Therefore, either mathematics do not apply to living organisms, or there must be some figures missing from the sum.

It seems to me that this book gives a clue to one at least of the possible missing figures, and I hope our scientists will give it the study it deserves. Mankind is at the last frontier. There is no new soil to be had in the horizontal plane. His hope lies in building new soil vertically.

Dr. Barrett asserts that by harnessing the earthworm in the way he recommends, the kitchen waste or garbage from a household of two or three members will furnish enough ideal earthworm food to breed tens of thousands of the soil builders each year. He himself, by this method, produces all the fruit and vegetables he and his family can eat, from one acre of land, as well as growing flowers and lawns. He discovered, in an experiment in his culture beds, that an acre of soil, if provided with enough organic matter, could support an astronomical population of earthworms.

He also quotes the Oregon State and New York State College of Forestry field studies as indicating that an earthworm population of from 250,000 to 1,500,000 an acre is enough to keep the earth as fertile and productive as man can want it.

In another publication, Dr. Barrett sums the matter up:

> "The problem facing civilization today is rebuilding the soil and restoring the earth to a form immediately usable for food production. By the slow process of nature, it takes 500 to 1,000 years to lay down an inch of topsoil. Under favorable conditions, a task-force of earthworms can do the same job in five years. An individual working with a lug-box or a compost pile can start building topsoil for his garden. A farmer working with a manure pile can do it with his farm.

154

A community utilizing a garbage dump can do it, or
a nation working with its resources can do it."

In connection with my work for the Soil Association, I have lectured on world soil erosion and the imperative need to restore, maintain and if possible increase the vitality of what soil is left. People often say, 'I realize the situation is appalling, but what can I do?' I feel this book at last contains a practical answer to that question. 'Feed earthworms.' This answer may sound flippant. I don't think you will think so when you have read this book. The technique is easy, and involves much less work than ordinary compost-making.

In all seriousness, I suggest that if everyone turned his attention to increasing the earthworm population (and there is no one who cannot do this, for it can be done even in a flower-pot or window-box) it might be the key to the survival of the human race. Through utilizing all organic wastes to feed earthworms and then deliberately putting them to work in the manner here described, it might be possible not only vastly to increase the fertility and productivity of the land already under cultivation, but also to arrest the further advance of deserts and dustbowls. This would give humanity a breathing space in which to learn how to bring other creative forces into play, so that water and life and the capacity to sustain vegetation may ultimately be restored to all the man-made deserts of the earth.

—*Eve Balfour*

On Earthworms

by Sir Albert Howard

▟

*Introduction to "The Formation of Vegetable Mould
through the Action of Worms with Observations on
their Habits" by Charles Darwin.*

When I first learned from the publishers of their intention
to reprint Charles Darwin's *The Formation of Vegetable
Mould* as one of their contributions to the discussion of the
principles underlying farming and gardening, the timeliness
of this step needed no argument. Present-day agricultural and
horticultural teaching and research are being critically examined
with a view to their speedy reform. Only good could result from
the re-publication of the results of some forty years' observation,
experiment, and thought devoted by our greatest naturalist to the
part played by earthworms in the history of the world. And, in
particular, to the manner in which they prepare the ground for
the growth of plants and of seedlings of all kinds.

When this intimation was followed by a request to write an
introduction which would link up Darwin's work with the recent
results on the earthworm and with the present controversy about
artificially stimulated crops, I at once agreed.

For some time, I had been seeking for the most effective and
convincing foundation on which the reformed agriculture of the
West could be based. And, at the same time, which would direct
future agricultural research into biological channels. This is an
ideal starting point and one which would call attention to the role

of the unseen labour force of the soil and to factors which the advocates of chemical farming could not lightly dismiss. No more effective basis for the organic farming and gardening of tomorrow could be found than the long and painstaking investigations described in this volume. It would help to accomplish for the temperate regions what King's *Farmers of Forty Centuries* has already done for the tropics and subtropics.

If this book, which deals with the formation of vegetable mould through the action of earthworms, is the real foundation for the study of the principles underlying farming and gardening, why should the renewed study of such a well-known work be necessary today? Why did the book fail to influence agricultural teaching and research when it first appeared in 1881?

The answer to these questions will be found in the manner in which investigations on farming have developed since 1840, when Liebig's *Chemistry in its Applications to Agriculture and Physiology* was published. This essay, coupled with the results of the well-known wheat experiments on the Broadbalk field at Rothamsted, exercised a profound influence on the minds of investigators and farmers alike. Till about the end of the nineteenth century, agricultural science was still a branch of chemistry. *The Formation of Vegetable Mould* through the Action of Worms appeared when the ideas founded on the Liebig and Rothamsted traditions were at their zenith.

It occurred to no one that the manner in which earthworms (1) periodically expose the surface soil to the air, (2) sift it so that no stones larger than the particles they can swallow are left in it, and (3) at the same time mingle the mould intimately together like a gardener who prepares fine soil for his choicest plants—were all matters of the most profound significance in crop production. The attention of all concerned was directed solely to the chemistry of the soil water, i.e. to a single factor only of a vast biological complex.

That Darwin's work was without influence on agricultural education in the nineties I know from personal experience. It was during the academic year of 1896-7 that I attended the various courses of lectures necessary to obtain the University Diploma of Agriculture at Cambridge. No references to earthworms and their work in preparing the food materials needed by crops were made in the lectures on agricultural chemistry offered. I never even heard Darwin's name mentioned in connection with the maintenance of soil fertility, but I was told much of the virtues of artificial manures for increasing crops and of the efficacy of poison sprays for controlling plant diseases.

It was not till the next two academic years, 1897-9, when studying for Parts I and II of the Natural Sciences Tripos that I first came into intimate contact with Darwin's books, including the one dealing with his studies of the earthworm. Later, during my Indian service (1905-31) when I became deeply interested in the preparation of humus on a large scale, the place of the earthworm and of the termite in converting crude vegetable matter into food materials for the crop appeared more and more important as the years passed.

It was after my retirement from service in India in 1931, when steps were taken to bring the large-scale preparation of humus to the notice of the farming community. I began to realize the urgent necessity of preserving the earthworms in the farms and gardens of Great Britain from destruction. Case after case came under my observation where the continued application of artificial manures and the use of poison sprays like lime sulphur and tar oils for keeping the pests of fruit trees in check led to the destruction of earthworms on a colossal scale.

Eventually, during the years 1935-8, I was able to pay the closest attention to the way artificial manures either reduce or eliminate altogether the earthworm population.

On some 3,000 acres of land near Spalding in south Lincolnshire, the property of the late Mr. George Caudwell, I spent many weeks in working out an improved method of green-manuring for the potato crop and was given every possible facility and help by my host in getting to the bottom of potato growing on the alluvial soils of Holbeach Marsh.

On Mr. Caudwell's farms about 1,500 acres of potatoes were raised every year on artificials only. For a time, exceptionally heavy crops had been obtained. But by 1935, the soil was showing distinct signs of wearing out and steps had to be taken to increase the content of organic matter. My opinion was asked as to the best method of doing this. I recommended a return to mixed farming which should include laying down a few hundred acres every year to temporary leys and the return to the old East Anglian custom of mucking the second clover crop before plowing under in the late summer.

This, however, would have involved a considerable reduction in the area under potatoes, to allow for the maintenance of cattle and pigs and the production of large quantities of farmyard manure. Unfortunately, both the beef and the pig trade at that time were under a cloud and Mr. Caudwell considered my proposals would not pay at the prices of meat then ruling.

He pressed me to suggest some alternative for increasing the organic content of his land, making the best use of about 1,000 tons of farmyard manure produced by his team of some hundred horses. This manure was therefore used to activate a green-manure crop of beans raised on the land under peas, which were always grown between two potato crops. The peas were cut in early July, carried to the shelling machines, the pulpy residues being returned immediately to the surface of the same land which, in the brief interval, had been plowed and sown with beans. On this layer of pea haulm pulp, a moderate dressing of farmyard manure (about five tons to the acre) was immediately spread, so that the newly sown bean land was covered with a double layer of organic matter—pea haulm pulp below and farmyard manure above. This provided the raw materials needed to manufacture a thin coating of humus on the surface of the soil under the growing beans. The beans soon grew through this fermenting layer, which the young foliage kept moist.

By October, the beans were in flower and from three to four

feet high. They were lightly turned under, together with the layer of humus which had, in the meantime, formed on the surface. The green crop was in this way provided with a very efficient activator as well as ample oxygen for conversion into humus. Nitrification took place before the next crop of potatoes was planted the following year. This method of green-manuring worked well and the tilth markedly improved. But, besides this dressing of organic matter, Mr. Caudwell insisted on applying about 15 cwt. to the acre of a complete artificial manure containing a large quantity of ammonia sulphate. To keep blight at bay, the potatoes were frequently dusted with copper salts.

The effect of all this was a heavy crop of potatoes, but the exceedingly sparse earthworm population did not increase as it might have done had no chemicals been used. Here was an excellent opportunity of observing the effect of humus supplemented by artificials on the earthworm population. These animals found such soil conditions unsuitable and no obvious increase in their numbers occurred.

This agreed with another observation of some interest, which was frequently repeated. In following the plows in the autumn and spring in the Spalding area, I always found that where heavy dressings of artificials were used every year, with or without organic matter, earthworms were rare. I sometimes walked half a mile after the plows and cultivators without seeing one. Further, on this artificially manured land the dense flocks of seabirds which so often follow the plow were seldom to be seen. These birds had evidently learned from experience that certain areas of Holbeach Marsh were not worthwhile as feeding grounds.

In the course of these green-manure experiments in Lincolnshire, I spent some time in studying the earthworms on land similar to Mr. Caudwell's farms but which were regularly dressed with farmyard manure. Here, earthworms were abundant and in some of the old tunnels I frequently observed the reaction of the roots of the potato (King Edward) to fresh worm casts. The fine roots often followed these tunnels downwards, but whenever they passed the earthworm casts, a fine network of roots was given off laterally which penetrated the casts in all directions. Obviously, the potato was making full use of these accumulations.

Why do the roots of the potato always invade earthworm casts? The answer is to be found in the work of a number of investigators.

In 1890, Wollny began a study of the place of earthworms in agriculture (*Forschungen auf der Gebiet der Agrikultur-Physik*, 13, 1890, s.381). He found, as a result of five years' work with cereals and legumes, that the mere addition of earthworms to soil led to a marked increase of grain (35 to 50 percent) and of straw (40 percent) above that of similar cultures without earthworms. Equally favorable results were obtained with flax, potatoes, and beetroots. In not a single instance did his cultures suffer any damage from earthworms. No support was, therefore, obtained for the common assertion that earthworms pull up young plants and carry them into the mouths of their burrows.

Having seen for himself that soil containing earthworms was considerably more fertile than a soil free from these animals, Wollny set to work to ascertain the cause. He found that earthworms markedly improved the permeability of soils and led to better aeration. With regards to the chemical composition of the soil inhabited by earthworms, he observed a considerable increase in the soluble nitrogen and in the available minerals as compared with similar worm-free soil. It was not determined if these results were due to worm casts or to the dead earthworms.

Wollny's work was afterwards confirmed by a number of investigators. In 1910, Russell (*Journal of Agricultural Science*, 3, 1910, p. 246) showed that earthworms contain 1.5 to 2 percent of nitrogen and decompose rapidly and completely, thus providing plant food to the soil in which they die.

In 1942, H. J. M. Jacobson of the Connecticut Experiment Station compared the composition of the earthworm casts and uncontaminated soil from the farm of Christopher M. Gallup at North Stonington and obtained the following results:

	Lb. per Million of Soil		Ratio of castings to soil
	Casts	Soil 0-6 in.	
Nitrate nitrogen	22	4.5	4.89
Available phosphate	150	20.8	7.21
Replaceable potassium	357.8	31.9	11.22
Humus	89,500	57,800	1.55

Curtis has summed up these interesting results in a press note issued by the Connecticut Station, which was reproduced in the *Gardeners' Chronicle* of 17th July 1943. These analytical results amplify and re-state in terms of chemistry Darwin's conclusion that:

> 'worms prepare the ground in an excellent manner for the growth of fibrous-rooted plants and for seedlings of all kinds.'

They also direct attention to a sadly neglected branch of the chemistry of the soil—the part played by the waste products of the soil population in plant nutrition. A detailed account of these Connecticut investigations has been published in *Soil Science* (58, 1944, pp. 367-75).

How does a dressing of a complete artificial manure affect the rest of the soil's animal population? It is a matter of common knowledge that chemical manures influence the number of animals like moles and toads. Both these prey on earthworms. It is well known that moles do not work in land which has been treated with chemicals. An interesting confirmation of this fact was recorded in an account of some experiments on grassland in Scotland early this century.

Between plots manured with farmyard manure strips were treated with a complete artificial. It was observed that the moles, which were abundant in the two dunged plots, always moved from one of these plots to the other in a straight line across the artificially manured plot and never threw up any molehills on this land. They seemed to know by instinct that such areas were useless as feeding grounds. From such a simple experiment as this much could be learned. The artificials depleted the earthworm population and this could easily be demonstrated by a set of counts. The reaction of moles to artificials would be obvious from the absence of molehills. If the behaviour of the grazing animal is watched, it will be found that the two dunged plots will be grazed to the ground, while the artificially manured plots are only lightly picked over. When, in the cycle—humus-filled soil, the grass carpet, and the animal—a substitute for humus is introduced in the shape of chemicals, the earthworms first abandon such soil and the moles follow suit. The grazing animal records an adverse verdict on the quality of the forage.

These results have been confirmed in the United States where ammonium sulphate is sometimes used to destroy the earthworms on golf putting greens and tennis lawns because the castings interfere with these games by clinging to the ball.

In *Farmers' Bulletin 1569* of the United States Department of Agriculture published in 1935 it is stated:

> 'The results of three years' application of ammonium sulphate to sod on the experimental farm of the Department of Agriculture at Arlington, Pa., for fertilizing purposes have shown incidentally that earthworms were eliminated from the plots where this chemical was used. When applied to soils which are naturally neutral or slightly acid in character, this fertilizer creates a strongly acid condition that is distasteful to the worms and they disappear.'

The use of artificial manures is not the only modern practice which destroys the earthworm. Hardly less injurious are the poison sprays such as Bordeaux mixture and other powders containing copper salts, tar oils, and the lime sulphur washes used for the control of insect pests. Perhaps the most complete account of such results is that given by Dreidax in the *Archiv für Pflanzenbau*, 7, 1921, and in *Rationelle Landwirtschaft*, Wilhelm Andermann, Berlin, 1927. The first of these papers concludes with a long list of references dealing, among other matters, with investigations on the earthworm since *The Formation of Vegetable Mould* was published in 1881. In the latter work, Dreidax sums up his observations on the deleterious effects of poison sprays on the earthworm population of vineyards in Germany.

While examining a vineyard situated in the Markgraf near Auggen in South Baden, in which poison sprays were constantly used, he found a surprising fall from the abundant earthworm population of meadows and orchards under grass adjoining the vineyard to that of the vineyard itself. The growth of the vines always corresponded closely with the number of earthworms. The rows next to the grass border were well developed. The vines in the centre of the vineyard, where there were no earthworms, did badly. In this investigation, the earthworm population under the grass border invariably stopped dead wherever poison sprays reached the turf.

A visit to almost any orchard in Kent during the spring immediately after the trees are sprayed with tar oils or lime sulphur will be sufficient to prove how harmful this spraying is to the earthworm population. The ground soon afterwards is covered with a carpet of dead worms.

It is in the United States of America that Darwin's studies of the earthworm have attracted most attention and have also been carried further in a number of useful directions. The reader interested in these developments should begin with the work of Dr. Oliver who, in 1937, published *Our Friend the Earthworm*, which contains a detailed account of his studies.

Starting life as a doctor, quite by chance he read Darwin's account of the work of the earthworm and at once began to investigate the relation of these animals to crop production. Soon he obtained evidence which fully confirmed Darwin's findings. Thus was started a series of experiments on the culture of earthworms, on the production of new types by hybridization, and the distribution of egg capsules for the purpose of re-stocking and improving derelict land almost devoid of these creatures. His success was immediate and in two years he sold his medical practice and set up as a landscape engineer. His main work was to improve private estates and public parks.

By 1920, he had become independent and was able to move to Los Angeles, where he pursued his investigations on a ten-acre experimental farm. In 1937, he published his results in the above-mentioned work of three volumes, which has helped hundreds of farmers all over the United States to restore fertility to barren land in which the earthworm population had been destroyed by artificial manures and poison sprays.

One factor in Oliver's career must be emphasized. His mind was prepared for the message contained in *The Formation of Vegetable Mould* by his early connection with farming, a detailed account of which will be found in the issue of *Organic Gardening* (The Rodale Press) of June 1943.

As a small boy, he went to live with his grandfather in Huron County, Ohio, on a 160-acre family holding which had been farmed continuously for sixty years on organic lines. This farm was divided into four blocks of 40 acres, one of which was taken up by the homestead, garden, orchard, and park, the other three blocks being used for mixed farming. In the centre of the whole 160 acres was the farmyard of 2 acres, which communicated directly with each of the four 40-acre blocks by large, swinging gates.

In the centre of the farmyard, in which the livestock were kept during the winter, was the compost pit, 50 feet wide and 100 feet long, which had been excavated to a depth of 2 feet. Down the middle line of this compost pit, about 20 feet from each end were two heavy posts about 12 feet high, each connected with the barn by a cable furnished with large travelling baskets, by which the manure from the great barn was transported each morning to the compost pit, where it was evenly spread. When necessary, the contents of the compost pit could be flooded by gravity flow from a neighbouring stream, which also supplied the drinking troughs of the livestock. The heart of this farm, on which crop failures were unknown, was the compost pit and its vast earthworm population of several millions.

An essential item of the daily work of the farm was the care of this compost pit. Every morning, the barn was cleared of the droppings of the livestock, which were evenly spread, together with all the available soiled litter, on the compost heap. When about a foot deep, several tons of red clay from the floor of a pond was distributed all over the compost pit. In this way, the vast earthworm population was supplied with organic and mineral food. After the spring thaw, the upper layer of the compost pit was removed and the rich dark crumbling layer of sweet-smelling worm casts was removed by shovels into wagons and spread on the fields just in front of the plows. In this way, an effective addition to the food materials needed by the new crop, as well as a copious supply of earthworms and egg capsules, was given to the land. Care was always taken to leave behind in the compost pit an adequate supply of earthworm castings, which acted 'as a 'mother substance' for the composting of the surplus top material when that was returned to the pit and composting was restarted.

This early experience naturally influenced Oliver in his later work. He had observed in his boyhood that earthworms will thrive and that a concentration of many thousands to the cubic yard is possible, provided a suitable environment, sufficient moisture, and ample food are supplied.

Another feature of this Ohio farm was the regular rotation—two straw crops (wheat and maize)—followed by a temporary ley of mixed timothy and clover. Every year, some forty acres of this ley were turned under. These areas always contained an unbelievable earthworm population.

A further detail must be mentioned. Four acres were in the orchard, while other fruit trees were planted in the hedgerows. Oliver's grandfather never allowed these trees to be cultivated. His motto was: 'Never disturb the soil under a tree. The earthworms are the best people for taking care of a tree and I don't want them disturbed.'

Oliver concludes the autobiographical fragment, from which the above account of his grandfather's farm is taken, with the following words:

> 'In this example of my grandfather's earthworm farm, we have the technique of utilizing the earthworm in general farming operations either on a large or on a small scale. From my experience as a small boy growing up on this farm with much friendly and loving instruction from my grandfather on the subject of earthworms, and in my own work covering a period of more than forty years, I am fully convinced that the eventual salvation of the soil of our country will include the harnessing of the earthworm as one of the major measures. And from my experience I know that the soil can be made to produce several times as much food as the present average through the proper harnessing and utilization of the earthworm under control.'

The preceding paragraph will help to explain the origin of the present-day interest in the place of the earthworm in farming, which is rapidly growing in the United States of America and which is now linked up with the campaign against artificial manures and poison sprays.

There is a growing volume of evidence from all over the world that agriculture took the wrong road when artificial manures were introduced to stimulate crop production and when poison sprays became common to check insect and fungous pests. Both these agencies destroy the earthworm and thus deprive the farmer of an important member of his unpaid labour force. There is also a strong case for believing that one of the roots of present-day disease in crops, livestock, and mankind can be traced to an impoverished soil and that these troubles are aggravated by the use of chemical manures.

The publication of the evidence which indicates that all is not well in our farming and gardening has very naturally disturbed the advocates of chemical farming. What has been described as the war in the soil has broken out and is now in full swing. The most effective way for all concerned of conducting this contest will be to pose to Mother Earth herself the question: What is your decision in this battle between organic and inorganic manuring? When this decision has been duly given, it will have to be interpreted. Livestock can be relied upon to say that the grain, fodder and forage raised with humus is far superior to that obtained with the help of chemicals. The earthworms will unhesitatingly plump for organic manuring. We can read their message by a simple count and by observing their general condition and activities.

There is no better soil analyst than the lowly earthworm. Our most experienced gardeners invariably judge the condition of their plots by the earthworm content. If in the autumn cultivation one large well-fed active lob-worm is turned up with each spadeful of soil, they consider that their land is in excellent condition for the next year's crop.

If, on the other hand, earthworms are few in number, pale in colour, and rolled up in a ball, they consider that a good dressing of organic manure is needed (King, F. C., *Gardening with Compost*, Faber and Faber, 1944, pp. 70-76). All this agrees with the way the tribesmen in the Sahara judge the fertility of the soils of the oases by the number of the earthworm casts. These, as we have seen, constitute the perfect food for plants. Obviously, we should do all in our power to increase this supply by providing the earthworms with the food and with the working conditions they need.

In directing attention to one of Nature's chief agents for restoring and maintaining the fertility of our soils, the publication of this new edition of Darwin's book will do much to establish the truth that Nature is the supreme farmer and gardener. The study of her ways will provide us with the one thing we need—sound and reliable direction.

—*Albert Howard, 4th January, 1945.*

ALSO FROM
A DISTANT MIRROR

Bechamp or Pasteur?

A Lost Chapter in the History of Biology

by Ethel D. Hume

352 pages

Hardcover, paperback, ebook

This volume contains new editions of two titles which have been available only sporadically in the decades since their publication.

R. Pearson's *Pasteur: Plagiarist, Imposter* was originally published in 1942, and is a succinct introduction to both Louis Pasteur and Antoine Bechamp, and the reasons behind the troubled relationship that they shared for their entire working lives.

Whereas Pearson's work is a valuable introduction to an often complex topic, it is Ethel Douglas Hume's expansive and well-documented *Bechamp or Pasteur? A Lost Chapter in the History of Biology* that covers the main points of contention between Bechamp and Pasteur in depth sufficient to satisfy any degree of scientific or historical scrutiny. It contains, wherever possible, detailed references to the source material and supporting evidence.

No claim in Ms Hume's book is undocumented. The reader will soon discover that neither Mr Pearson nor Ms Hume could ever be called fans of Pasteur or his version of science. They both declare their intentions openly; that they wish to contribute to the undoing of a great medical and scientific fraud.

A DISTANT MIRROR

WWW. ADISTANTMIRROR. COM. AU

EUGÈNE MARAIS

THE SOUL
OF THE APE
&
MY FRIENDS
THE BABOONS

Eugene Marais spent three years living in the South African wilderness in close daily contact with a troop of baboons. He later described this as the happiest, most content time of his troubled life. This period produced two works which are testament to his research and conclusions.

Firstly, there was a series of articles written in Afrikaans for a newspaper that were published in an English translation in 1939 under the title *My Friends the Baboons*. These pieces were written in a popular vein suitable to a newspaper readership. They are a journal; a series of anecdotes and impressions.

The Soul of the Ape, which Marais wrote in wonderfully clear and precise English, was the more serious scientific document; however after his death in 1936, it could not be found. It was lost for 32 years, and was recovered in 1968, and published the following year.

The excellent introduction by Robert Ardrey that is included in this volume was part of the 1969 and subsequent editions of *The Soul of the Ape*, and adds greatly to an appreciation of its importance.

Together, these three texts give us as complete a picture as we will ever get of Marais' three year study of these complex relatives of humanity, and its implications for the study of consciousness.

It was always his intention that his two bodies of work, on termites and apes, were companion pieces in the search for an understanding of the psyche that would span the gulf between the insect and primate worlds. The point of Marais' work was, always, the mystery of consciousness itself, on which grounds it is as relevant as ever.

A DISTANT MIRROR

WWW.ADISTANTMIRROR.COM.AU

The Wheel of Health

by Dr Guy Wrench

182 pages

Paperback
ISBN 978-0980297669

"Why not research health, as well as disease?"

The Hunza of northern Pakistan were famous for their extraordinary vitality and health.

Dr Wrench argues that in part at least, this is because their food was not made 'sophisticated', by the artificial processes typically applied to modern processed food. How these processes affect our food is dealt with in great detail in this book.

The answer that Dr Wrench uncovered in his researches goes deeper than just the food, though. The real answer lies in what was special about the Hunza's water supply.

Contents

1 – The Hunza People 2 – A Revolution in Outlook

3 – The Shift to Experimental Science 4 – The Start

5 – Continuity and Heredity 6 – Other Whole-diet Experiments

7 – Fragmentation 8 – The Cause of Disease

9 – The Hunza Food 10 – The Cultivation of Hunza Food

11 – Progress by Return 12 – An Experiment

A DISTANT MIRROR
WWW.ADISTANTMIRROR.COM.AU

ADISTANTMIRROR.COM.AU

9 780648 859420